Advance Praise for

Trash Talk

"As this sprightly book makes clear, it's hard to throw something 'away' because a finite planet doesn't really have an away—not even in orbit, where space junk is piling up. So we better start thinking through some new approaches!"

—Bill McKibben, author of *The End of Nature*

"*Trash Talk* is a brilliant, revelatory delight to read. It is an anthropological whodunit that thoroughly describes how much we extract from the living world and transform it into the dead world. Who thought a book on trash could be a page-turner. Think again!"

—Paul Hawken, author of *Drawdown* and *Regeneration*

"Like all of Iris Gottlieb's books, *Trash Talk* is a marvel of fascinating information, relatable insights, and laugh-out-loud hilarious illustrations and commentary. You'll be engrossed, entertained, and horrified by all the waste-related facts packed into the book, along with genuinely useful details about the world of trash that most of us know so little about. Everyone should read this book!"

—Margaret (Mei) and Irene Li, authors of *Perfectly Good Food: A Totally Achievable Zero Waste Approach to Home Cooking*

TRASH TALK

An Eye-Opening Exploration
of Our Planet's Dirtiest Problem

IRIS GOTTLIEB

A TarcherPerigee Book

tarcherperigee

an imprint of Penguin Random House LLC
penguinrandomhouse.com

TarcherPerigee with tp colophon is a registered trademark of
Penguin Random House LLC

Most TarcherPerigee books are available at special quantity discounts for bulk purchase for sales promotions, premiums, fundraising, and educational needs. Special books or book excerpts also can be created to fit specific needs. For details, write SpecialMarkets@penguinrandomhouse.com.

Library of Congress Cataloging-in-Publication Data
Names: Gottlieb, Iris, author.
Title: Trash talk: an eye-opening exploration
of our planet's dirtiest problem / Iris Gottlieb.
Identifiers: LCCN 2023045926 (print) | LCCN 2023045927 (ebook) |
ISBN 9780593712771 (trade paperback) | ISBN 9780593712788 (epub)
Subjects: LCSH: Refuse and refuse disposal. |
Recycling (Waste, etc.) | Refuse collection.
Classification: LCC TD791.G718 2024 (print) | LCC TD791 (ebook) |
DDC 363.72/8—dc23/eng/20231129
LC record available at https://lccn.loc.gov/2023045926
LC ebook record available at https://lccn.loc.gov/2023045927

Printed in Canada
1   3   5   7   9   10   8   6   4   2

Book design by Shannon Nicole Plunkett

To the garbage collectors and
waste pickers of the world

CONTENTS

SO, WHAT IS TRASH?

By generic definition, trash is anything worthless, unwanted, or discarded.*

But what we deem *trash* can be drastically different from how our neighbors, other cities or towns, or countries across the world value those same objects. The classic (and overused) adage "one man's trash is another's treasure" is applicable beyond making yard art from scrap metal or scoring big on *Antiques Roadshow* from a thrifted painting. It's relevant on a personal, global, and economic scale. It illuminates that trash is not always unanimously categorized or similarly managed across the board; for some it's a treasure, for others it's invisible, and for more it's an enormous human and biological hazard. Humans don't agree on what makes trash *trash*. The way we individually and collectively determine worth and value is lacking consensus. Our discord is a reflection of the global shift in wealth, labor, skill, and resourcefulness.

Depending on who we are and where we live, our definitions of *worthless*, *unwanted*, or *discarded* might vary wildly. In highly consumeristic cultures, the amount of trash produced is much greater, in part because of access to products, disposable income, and single-use cradle-to-grave

---

* The words *garbage, trash, rubbish, waste,* and *refuse* are used interchangeably, but there are historical differences between the terms. The earliest appearance of the word *garbage* was in a recipe for a meal of chicken giblets and organs in *A Boke of Kokery,* a cookbook from fifteenth–century Britain. Historically, *garbage* has referred to organic matter: kitchen scraps, yard waste, or other "wet" materials. *Trash,* by contrast, is any "dry" refuse: paper, plastic, or cans. *Refuse* and *rubbish* are (very British) catchalls, which include wet, dry, and any other discarded items ranging from couches to cell phones to concrete. Because it's not common practice to use these words separately anymore, they will be used interchangeably throughout the book.

production. The United States is, no surprise, the world's biggest generator of household trash. Each American generates 1.38 tons of waste annually, contributing to the global trash production of 4.5 trillion pounds per year.[1] For context, that is 22.5 billion blue whales. Or if you were to stack it, it'd be 782,608 Great Pyramids of Egypt.[2] Or 42 times the total number of humans that have existed in the history of Earth (if each person was a pound of trash). *Every year.*

The production of trash is unfathomable, and yet our exposure to the amount of waste we generate is mostly from our own homes (or on the curb if you live in New York City) when we put it in a bin or toss it down the chute to be taken away in units of tied-up bags. We see what we create on a moment-to-moment basis, but not the cumulative effect of our neighbors, our cities, our countries. I see a bag of kitchen garbage fill up over the course of two weeks, add my bathroom trash to it, put it in the big trash can, and start over. I don't think about it much once it's out of the house.

Throughout many wealthy nations are many layers of bureaucracy or infrastructure that obscure our trash systems, accentuating our ignorance of the magnitude of waste. In some ways, this opacity serves a practical, beneficial purpose: the less we can see it, generally the further it is from our daily lives, and the less likely it is to cause serious sanitation and health issues. A nonexistent sanitation system contributed in part to the explosion of bubonic plague that wiped out more than twenty-five million people in the 1300s. When our infrastructures properly prevent biohazards that can cause illness, opacity serves the well-being of those who have the privilege to not be near mountains of garbage or incinerators. There are many, many populations throughout the United States and other wealthy countries that don't have the benefit of distance from trash and its hazards. I will delve more into environmental racism and the inequity around waste disposal later in the book.

That same level of opacity also allows corporations to avoid their responsibility to generate less waste and handle the disposal of what they produce. Because much of how we handle waste is hidden, these companies can advertise false promises of doing better ("greenwashing") and create self-generated environmental success metrics to keep the public in the dark, obscuring the true practices of big manufacturers and brands. If we don't see the scale of what's wrong, then we have very little ability to see what needs to be fixed beyond our small choices.

In many other places in the world, trash is anything but invisible. It is omnipresent in the form of the e-waste and recycling that the United States and Europe ship away—waste that provides dangerous and extremely low-paying jobs. Tons and tons of discarded clothing sits in the Atacama Desert in Chile. Avalanches of garbage pose life-threatening hazards in Jakarta's landfills. Makeshift villages are built atop landfills in Lagos, Nigeria.[3]

The truth is there are many different facets to trash: what it's made up of, how it's handled, and who is impacted by it—both on personal and global levels. This book covers the historical, psychological, geographical, environmental, and class inequities around trash; and it all comes down to how we determine the worth of objects and people.

I am not a virtuous saint of zero waste here to instruct you, reader, to *do better*. I sometimes know what's "right" and still don't do it. I have had a jar of dead batteries on the kitchen counter for two years that will most likely end up in the trash and cause a problem at the waste facility that I'll never witness, and I will feel guilty until I forget and the batteries accumulate once again—at least until I have nothing left that's battery operated. I don't know what to do with empty cans of spray paint. I usually throw away plastic peanut butter jars after I let my dog gnaw on them because the amount of water to get them slightly less greasy doesn't seem worth the trade-off for them to still potentially wind up being rejected by a recycling plant. I forget my reusable bags 50 percent of the time and usually don't buy in bulk. I have a dog who produces waste that then goes into a plastic bag and becomes more waste. I have a stack of broken cell phones with screens held together by packing tape in my office, but can't bring myself to go to the mall's Apple store because it's an all-around hellish experience. I subscribe to *Vanity Fair* to stay up to date on celebrity gossip.

These are very minor waste-producing behaviors in the grand scheme of things, and on the whole I unintentionally enact a lot of waste-reducing behaviors, which I'm afforded because of my life circumstances. Often using less costs more, and I have a certain level of class privilege that allows me to make choices that reduce waste. I live in a wealthy country, work from home, have access to a range of grocery stores, have the space to compost, drive a relatively efficient car, and have no children. I rarely (if ever) buy new clothes, usually shopping at thrift stores or wearing the same thing for ten years. I don't get takeout because I am rendered useless by the decision-making involved. Due to where I live and my class, race, and access to infrastructure, I have more choices available than many who must engage in waste production with much less safety and freedom.

Trash is inherently and sometimes confusingly imbalanced. The more wealth you have, the more options for wastefulness are at your disposal: frequently buying new things, traveling, getting takeout, not *having* to reuse. This is an intentional societal display of power through the culture of disposability.[4] But in the same way that the wealthy have access to buy and dispose of more, the options to buy less are greater as well. The opportunity to grow one's own food as a hobby requires free time; eco-friendly options are almost always more expensive than heavily packaged products; reusable water bottles assume one has access to clean water to refill them; and composting services are generally privately owned and charge collection fees.

For those who do not have much wealth, affordable and accessible food choices often come in excessive packaging, and neighborhoods struggle with uncollected trash as a public health issue. Cheap and poorly made goods are more affordable but end up costing more in the long haul to frequently replace—and this cycle of waste can contribute to more socioeconomic disempowerment. We don't all have the time (or interest) to join Greta Thunberg on a two-week boat trip to avoid a plane ride.

The intention of this book is not to elicit shame or guilt around your personal choices. While those emotions can sometimes be useful for inciting change, we are just as likely to dig our heels in, shy away from, or turn our back on the topic altogether when scolded. I often give an eye roll at lists of "10 Things You Can Do to Solve Climate Change," and I don't want this book to be put down because it's asking or instructing you to tackle things that feel pointless or beyond your scope of action. How we live is complicated. The choices we're given do not set us up for success, and the systems in place are intentionally confusing. Much of the literature and education around trash encourages personal change through bubbly, action-oriented behaviors without the information about *why* it's important (or useless) beyond the basic idea of climate change or virtue signaling.

Action-oriented behaviors can certainly be useful and function to assuage our eco-guilt (the guilt we have around not choosing the best, greenest option) but at the end of the day, the system is broken, and we as consumers have quite minimal power with our individual actions. A friend of mine has no municipal recycling or garbage pickup. When she took both types of waste to her local landfill, they had her dump both into the same pile, which would never be sorted. This is not to say that recycling never works or never happens, but knowing the realities of the infrastructure in place around you gives context for which actions do or do not help. This is by no means a get-out-of-jail-free card to take a nihilistic approach and no personal responsibility, even though it's a tempting route to lean toward. It's certainly easy to slip into that mindset, and I often do, but there's a balance of not totally throwing intentional actions out the window while recognizing we are each but blips in the grand scheme of waste generation.

Instead of shaming you or filling these pages with optimism, I hope to illuminate the real societal frameworks that, for the most part, we can't see, so as to better understand where we fit into them, how to shift them, and how we got so deep into this global crisis. These systems range from advertising to manufacturing to class structures to ancient piles of oyster shells to bags of vomit on the moon to the mafia. Trash touches almost every facet of our daily lives, whether we are aware of it or not, and knowing more about the depths of the issue can show us where we can shift our behavior, and where it's the responsibility of larger systems to shift in order to halt current patterns and prevent future crises.

Context always matters, and there are extremely varied opinions when it comes to garbage. I will do my best to shed light on these topics and include different viewpoints on how these problems should be addressed. This is a rapidly shifting field of thought, and I am not an expert; but if anything, I hope context helps us all empathize with those negatively impacted by these problems, understand how we got to where we are, be angry at the right people, and navigate our personal responsibility. The situation is bleak, and this book is not a greenwashed portrayal of what's going on. There are certainly bright spots as well as weird, gross, fascinating, and surprising facets of the garbage world, but it's often seeing the dark underbelly of an issue that shocks us into grasping the reality of it.

I hope what you learn will make you curious about the garbage all around you that you never noticed before.

[VERBATIM EXCERPT FROM A BOKE OF KOKERY]

How Big Are We Talking?

"Enough to fill an Olympic-size swimming pool" is a common unit of measurement, but I would hazard a guess that most people have not stood next to an Olympic-size swimming pool. And if you have, there's a good chance you didn't store it away in the mental measurements folder with quarts, gallons, cups, and tablespoons. It's a pool, not a measuring cup. If you're an Olympic swimmer, I'm sorry I won't be using your preferred metric of comparison.

A COMPARISON OF
USELESS MEASUREMENTS

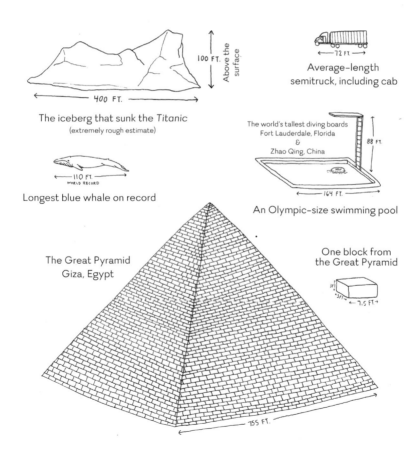

100 FT. Above the surface

400 FT.

The iceberg that sunk the *Titanic*
(extremely rough estimate)

72 FT

Average-length
semitruck, including cab

110 FT.
WORLD RECORD

Longest blue whale on record

The world's tallest diving boards
Fort Lauderdale, Florida
&
Zhao Qing, China

88 FT.

164 FT.

An Olympic-size swimming pool

The Great Pyramid
Giza, Egypt

One block from
the Great Pyramid

3 FT
3 FT
7.5 FT

755 FT.

The one pool-related moment that should be solidified in our collective cultural memory is the old woman swimming in spaghetti in *Patch Adams*, but that was a small pool in a backyard. The non-noodle-related Olympic pool is a common choice of measurement for garbage production—it holds eighty-eight thousand cubic feet. Useful? Not really. How about a blue whale, as I mentioned earlier, as a size comparison? Nope! The concept of one million? Not useful for most of us.

In an effort to find something that's an easily imaginable volume with which to quantify garbage production, let's try on a semitruck container for size. We see them all the time, and they're very big, but not so big as to be impossible to conceptualize. A standard semitrailer holds 3,800 cubic feet—twenty-three times less than an Olympic-size pool. So, those pools are really, really big.

Globally, humans produce eight hundred thousand Olympic-size pools of garbage a year—or 18.4 million semitruck containers worth of trash. We're back to the challenge of conceptualizing a million (much less eighteen of them), but it's easier to recognize the sheer volume of garbage when imagining 18.4 million semitrucks clogging the highways. At about seventy-two feet in length, that's 250,000 miles of truck. The National Highway System of the entire United States totals 161,000 miles, meaning every mile of highway in the US would be bumper-to-bumper semitrucks, with half of those being full in both directions.

Conceptualizing the scale of trash production is one of the hardest challenges of the issue—it's a scale too big for most of us to understand, and without really feeling grounded in the scope, it's hard to grasp the dire nature of the problem. Throughout the book there will be measurements of scale in tons, acres, and a portal to another dimension in Oscar the Grouch's trash can. While it's all mind-bending amounts, the bottom line is that it's enormous.

And next time you're at an Olympic pool (never), you can imagine it full of twenty-three semis.

A Note about This Book

Before truly delving into the world of trash, I want to give a preface about the object you're holding, including any e-readers out there. There is a certain level of hypocrisy involved in writing a book about trash, as it requires tons of paper, water, energy, chemicals, and ink to produce, and many copies of it will eventually end up in landfills (approximately 320 million books are tossed per year).[5] We aimed to print this book in the US on fully recycled paper, but due to opacity issues, concessions were made: printing in Canada using less environmentally friendly paper, though one that is certified by the Forest Stewardship Council. This book itself reflects the environmental tradeoffs ever present in consumer goods.

It's relatively difficult to find transparency around the true impact of publishing, in part because there are an enormous number of factors that go into any given publication, including the size of the publishing house, current supply and demand of resources, where the printing happens, how it's shipped, what's used to bind the spine, and if it's in color or not. Our increased demand for online shopping and instantaneous shipping from companies like Amazon has also made it more difficult for the publishing industry to make any meaningful progress toward more sustainable fulfillment and distribution methods. Beginning in 1995, Amazon began advertising itself as "Earth's Biggest Bookstore," and has held that title, currently making up over 60 percent of US book sales. Publishers heavily rely on the online retailer and have little control of how their books are handled once they leave the warehouse, despite hopes that the industry can be greener.[6] There is some pushback against the increasing monopoly of Amazon's bookselling within online shopping, with companies stepping into the ring. For example, Bookshop, a certified B Corp* online retailer of books from independent bookstores, aims to pull traffic away from Amazon to directly support local bookshops around the country.[7] Over 80 percent of their profits go directly to independent bookstores and they kept many afloat during COVID-19 shutdowns. If you love this book so much and want to buy a copy online for a rat-obsessed friend (see chapter 5) or your wasteful aunt, I encourage buying from Bookshop.org if possible.

On the production side of things: I will get more into paper further along, but it's a huge source of pollution, political conflict, and waste. Ac-

---

* According to bcorporation.net, B Corp certification designates "that a business is meeting high standards of verified performance, accountability, and transparency on factors from employee benefits and charitable giving to supply chain practices and input materials."

cording to *The Story of Stuff,* in 2010, unsold books accounted for "an average of 25–55% of what gets printed depending on the genre, which are usually either trashed, recycled, or sold to a discount bookstore."[8] They won't tell me how many unpurchased copies of my book are sitting somewhere awaiting their fate. The process of book manufacturing is toxic, energy intensive, and destructive, even though it has become less so with efforts to source sustainably harvested paper and use more biodegradable or recyclable binding glues. Every page in this book used an estimated three gallons of water to produce.[9] *Cradle to Cradle,* a book about sustainable design by Michael Braungart and William McDonough, took an innovative approach to avoiding the environmental cost of paper production by using recycled plastic as "paper" to bypass organic matter in favor of synthetic material that can be upcycled again and again. The book is even waterproof, if you want to do a little research in the tub. The technology isn't popular, but it has potential to serve as inspiration for publishers to increase efforts of reducing waste.

Every year, thirty million trees are used to print books just for the United States, which accounts for the largest number of books published per year globally, but it's certainly far from the entirety of the publishing world's mark on the environment.[10] In 2018, the World Intellectual Property Organization and International Publishers Association collected data from sixty-three countries/territories about book sales and printing, finding that the United States topped the chart with 2,597 million copies sold (remember, a huge number of printed books are not sold) followed by the UK at 652 million. In 2018, the US had 3.5 million registered ISBNs (the global system of numbers used to identify books), which dwarfs the UK at 185,721, Germany at 139,940, and Italy at 137,397.[11]

Penguin Random House, this book's publisher, is aiming to shift toward a more sustainable model of production, using 100 percent Forest Stewardship Council (FSC) certified harvested paper. In its US warehouses, which ship out 1.2 million books per day, there is an effort to reduce packaging with more efficient systems and less plastic wrapping. The US warehouses have offset their electric use with the purchase of renewable energy credits (a tricky and oftentimes smoke-and-mirrors solution, so take that with a grain of salt).[12] In a 2018 report, the Rainforest Action Network had a very different take on Penguin's efforts to make environmentally friendly moves, and determined that there was a serious need to catch up to the pack on sustainable accountability and action.[13]

One example of successful, transparent efforts toward sustainability is Hachette Livre, a French publishing company that very clearly states their practices and statistics. With a commitment to use sustainably, legally harvested paper, 91.5 percent of their books in 2018 were printed on FSC paper, and their distribution centers are equipped with balers and shredders to do in-house recycling rather than shipping their recycling to an offsite facility.[14] It's excellent and important that companies like Hachette are working toward a net neutral or net positive effect on waste production, and their transparency is equally important. Sharing honest reports about internal successes and flaws creates accountability when a company's dirty laundry is aired.

Books are valuable tools for learning and entertainment, and I am by no means arguing against the printing and selling of physical books. They're wonderful and important and how I make a living. Buying paper books is something that I love doing and do often. I'm historically a very slow reader and I hated pleasure reading until my twenties, so having a full bookshelf brings me a sense of satisfaction—making up for all those grumpy years of not reading. I get most of my books secondhand, many of which come from online secondhand sellers, Bookshop.org, Little Free Libraries, or thrift stores. I (sometimes) loan out books to trustworthy friends who won't steal them forever, and I keep a list of who has what—my own self-operated library.

Book printing and distribution practices are often heavily waste generating, but there is plenty of room for improvement with the technology and infrastructure there to accomplish it. Although consumer choice often matters less than we'd like it to, book buying is one realm in which individual choices can actually make an impact. Buying from Amazon hurts the industry with unnecessary packaging (a book surrounded by plastic in a box ten times bigger than the book), unrealistic shipping expectations that make transportation inefficient, and reduced customer bases for independent, local booksellers.

For those who prefer to read digitally, how does your e-reader stack up to paper books? At first glance, it might seem like they are more eco-friendly, as no trees were harmed in the making, and your library can be infinitely full without producing anything physical—no glue or ink or forests involved. *The New York Times* put together a step-by-step journey of e-reader production to compare its material and energy consumption to paper books.[15]

Here's how the numbers shake out:

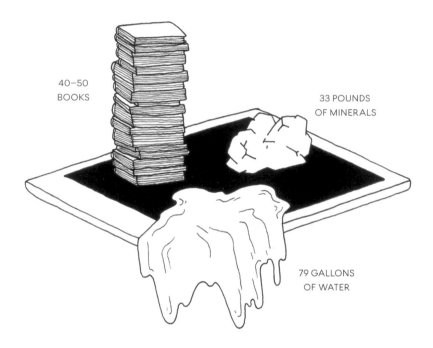

40–50
BOOKS

33 POUNDS
OF MINERALS

79 GALLONS
OF WATER

What's required to produce one e-reader:

- Thirty-three pounds of minerals—some toxic and some benign, but still energy consumptive to mine and refine

- Seventy-nine gallons of water

- One hundred kilowatt-hours of energy (generally fossil fuels)

- Sixty-six pounds of CO_2 emissions

- Seventy times greater adverse human health effects than paper book production

Forty to fifty books need to be e-read to even out the fossil fuel, water, and mineral usage of e-reader production with the production of paper books, and one hundred books need to be e-read to even out the carbon emissions of e-reader production.

To produce one paper book requires:

- Two-thirds of a pound of minerals

- Two gallons of water*

- Two kilowatt-hours of energy

- One hundred times fewer greenhouse gas emissions compared to e-readers

A landfilled paper book will release twice the amount of emissions (as methane and other greenhouse gases) in the landfill compared to the emissions generated to produce the book itself.

So, if you're a heavy reader, an e-reader might be a better option than buying *new* printed books. However, the most ideal options are to buy used books from local thrift stores or bookstores, or utilize your local library—both of which require little to no environmental repercussions, outside of transport to and from the locations.

If you no longer want this book (no hard feelings), consider giving it to a friend or donating it to an organization, library, or used bookstore.

---

* This is to make the pulp slurry that is pressed into paper. Depending on how water consumption is measured, some sources measure up to three gallons of water per sheet of paper.

Part 1

OUR TRASHY HISTORY

Chapter 1

ANCIENT SYSTEMS

"Probably from the moment when Adam and Eve were finished
eating their apple and wondered what to do with the core, we
have wrestled with what to do with our refuse."

—Larry VanderLeest, *Garbio: Stories of Chicago, Its Garbage,*
and the Dutchmen Who Picked It Up

We have what we have on Earth—it's a finite, closed system, and all the materials we are going to have are already here, so the thought exercise of how these materials have physically and psychologically shifted into what we toss out is an interesting one. Where was all the material before it got to a landfill? At what point does something lose its association with the natural materials used to create it and become easier to discard? Once something is printed on paper? Once ceramics are glazed with wild colors? Once it's hard to tell if furniture is made of wood or plastic imitating wood? How do we decide what trash is a metric of human success and what trash is a metric of human failure?

Writing this book made me think about museums full of artifacts from ancient peoples around the world—Egyptian sarcophagi from the 2600s BCE, Chinese glazed pottery dating back to 1600 BCE, Mesopotamian glass objects from 4000 BCE, the sunken Dokos ship from ancient Greece resting on the ocean floor since 2700 BCE. It brings the question: What defines trash? Must it be intentionally discarded, or can it also be what is left behind, willfully or not?

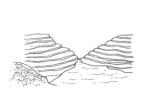

41,000 BCE
NGWENYA MINE
ESWATINI

2300 BCE
OLDEST MUMMY
EDFU, EGYPT

1600 BCE
GLAZED POTTERY
SHANG DYNASTY, CHINA

1ST CENTURY CE
MONTE TESTACCIO
ROME, ITALY

2700 BCE
DOKOS SHIP
BOTTOM OF THE AEGEAN SEA

6 BCE/IMAGINARY
SHIP OF THESEUS
YOUR THOUGHTS, YOU

9 CE
TOILETS
MAYAN RUINS

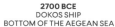

1925
PHOEBUS CARTEL
GENEVA, SWITZERLAND

1957
SPUTNIK
OUTER SPACE

1972
THE BLUE MARBLE
APOLLO 17, OUTER SPACE

1978
BEAN V. SOUTHWESTERN
WASTE MANAGEMENT CORP.
HOUSTON, TEXAS

1946
NYLON STOCKINGS VENDING MACHINE
NEW YORK, NEW YORK

1970S
EARTHSHIP MOVEMENT
TAOS, NEW MEXICO

1973
CANE TOAD
AUSTRALIA

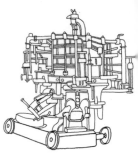

79 CE
POMPEII, ITALY

1346–1353
VECTOR OF THE BLACK DEATH PLAGUE
ASIA, EURASIA, EUROPE, NORTHERN AFRICA

1870
FIRST PLASTIC

1903
MECHANIZED BOTTLE MAKER
TOLEDO, OHIO

16TH CENTURY
KINTSUGI POTTERY
JAPAN

1776
KING GEORGE III STATUE
TOPPLED TO MAKE 42,000 BULLETS
BOWLING GREEN, NEW YORK

1901–PRESENT
CENTENNIAL LIGHT BULB
LIVERMORE, CALIFORNIA

2000
NOKIA 3310

2007
1ST GENERATION IPHONE

1987
MOBRO 4000 TRASH BARGE
LONG ISLAND, NEW YORK

2001
DEBRIS FROM 9/11 ATTACK
FRESH KILLS LANDFILL, NEW YORK

2021
TRASHED AMAZON RETURNS
DUNFERMLINE, SCOTLAND

When we think about ancient materials, we usually think of them as natural, from the earth. And if they're from the earth, they must easily be returned to it by decomposition, leaving little evidence behind after a period of time. Clay remains clay, stone remains stone, metal remains metal. But the process of extraction and transformation can completely change the properties of those materials and objects, rendering them permanent artifacts rather than circular resources in an ecosystem. Glass is a good example. When glass is made by humans rather than naturally by volcanoes (obsidian) or lightning, it's formed by heating silica-rich sand to a liquefying temperature of 3,090°F to produce a chemical transformation that can never be reversed. I would hazard a guess that almost no one thinks of sand as garbage, but once it changes forms it instantly becomes a manufactured, nonbiodegradable material. It is estimated that it takes one million years to degrade, but it could potentially take much longer (if not infinity) to break down, as no person has been alive for one million years to document the process.

If eyes are the windows to the soul, then trash is the window to the past. Some of the oldest evidence of human activity, ritual, innovation, diet, and population size of a community comes in the form of middens, or ancient garbage dumps. Humans are unique in our linear production of waste; in all other biological systems there is a symbiotic relationship among organisms to take care of waste. Scavengers feed from picked-over carcasses, feces decompose and serve as fertilizer, a squirrel's abandoned acorns sprout oak trees. At a certain point, humans tipped out of the natural rhythm between cyclical waste creation and the environment's use for it. Our garbage is at a peak moment of crisis, but "our species faced its first garbage crisis when human beings became sedentary animals."[1] Any time we settle somewhere, we leave remnants of our existence there. From early arrowheads and tools to pottery shards and glass, it turns out plastics aren't the only permanent markers of our propensity to create and discard.

Middens can be found almost everywhere that humans have congregated over the course of history, and exist today in the form of carbon-rich deposits from organic matter or collections of discarded tools and vessels. These sites are often recognizable because the layers of earth are darkened from carbon-rich material decomposition—food scraps, animal parts, or human burials.[2] I haven't ever rotated my compost and I've only added a couple pathetic handfuls of dried leaves in the past two years, so I have a pretty good sense of what the rich dark sludge left behind for millennia would look like.

Shell Mounds

One of the most common types of middens are shell mounds: huge deposits of shells found along the coast that are evidence of an abundance of shellfish available to coastal peoples thousands of years ago. These shell mounds often give insight about the sustainability of indigenous people's harvesting practices, as shown by the consistently large size of the bivalves (generally oysters)—evidence that they were allowed to grow to maturity and were not collected prematurely, preventing overharvesting.[3]

Shell mounds can be found around the world on coastlines, and due to the high alkalinity of shells, they have an amazing ability to preserve objects (both organic and not) lodged inside, providing a clear window into the past behaviors of the people who created the mounds. Sometimes shell mounds were used as burial sites or places of ritual, and often were big enough to define coastal landscapes. Hunt Revell, cofounder of Shell to Shore, explains that "humans have been gathering around oysters for ten to twenty thousand years, and the idea of the gathering place being the trash heap means this is one place people stayed for a while."[4] A mound that was demolished in Emeryville, California, was 30 feet tall by 350 feet wide,[5] and a preserved Japanese mound in Miyatojima is 2,600 feet long and 650 feet wide.[6]

Unfortunately, a huge portion of shell mounds around the world have been destroyed by erosion, colonization, and human development. These archaeological sites are extremely important to gaining understanding of the people who lived there, and they hold importance today to the tribes and populations who still live on those lands. There is a push to preserve some of the remaining middens in an effort to learn about ways of sustainable shellfish harvesting and the histories of populations that no longer exist, and to give ownership back to those whose ancestors lived in those places. On discussing the narrative that the earliest humans "chased the big boy mammoth across the Bering Strait from Russia to Alaska and spread across the modern day US," Revell shares findings that "some of these shell mounds on the West Coast from Washington to Chile have been around longer than the Clovis arrowheads found in Mammoth Park,"[7] meaning that shell-mound-creating populations potentially predated the popular notion that people first came to the modern-day United States over the Bering Strait.

Ancient indigenous groups might have set large rocks out into the ocean to cultivate oyster beds at a distance such that the bivalves were reachable and wouldn't wash out with the tide, but far enough to be in their ideal growing environments—thus creating "little gardens on the coast." Evidence of such ancient sustainable marine farming practices might indicate, Revell suspects, that these groups "weren't as nomadic as we think, because it's a really sustainable food source rather than you gotta go chase the mammoth and feed your family for the year."[8]

Shell mounds provide clues to both cultural and ecological practices, but there are other planned waste systems that can be found around the world that uniquely illuminate social patterns, city planning, and timelines of a city's rise and fall. The ancient Maya had garbage and recycling systems, and the city of Troy was built upon layers and layers of its own debris.[9] There is evidence of landfills and recycling in the city of Elusa dating back to earlier than 550 BCE, in what is now the

WHAT HAPPENS TO OUR SHELLS TODAY?

While the destroyed middens cannot be restored, there is work being done to recycle oyster shells to rehab depleted oyster beds, build infrastructure to mitigate erosion, and encourage healthy oyster populations for ecological health and water quality. When Revell started Shell to Shore, an oyster recycling and marine ecosystem rehabilitation organization based in Athens, Georgia, his goal was to tackle systemic flaws and explore possibilities of using recycled oyster shells to rehab coastal ecosystems and reduce waste. Revell's belief is that an oyster shell "isn't trash, it's a self-sustaining organism. This shell is not alive, but it's part of what creates oyster reefs. Not only that—it can be repurposed for all kinds of other things, so there is absolutely no reason this should be going into a trash can."[10]

Oysters are a good example of the global food system creating problems that both serve and harm the environment and local populations. Shipping food (in theory) allows for products to reach any corner of the globe, some of which is necessary to feed populations with little food access. But sometimes it is motivated by unnecessary decadence—like oysters in Nebraska. This doesn't serve to alleviate food scarcity, and by virtue of its habitat, oysters can never be local unless the ocean swallows about fifteen hundred miles of the West Coast. To accommodate the excessive demand for a slow-growing species, farmed oysters have become a huge business, much like many species of fish and shrimp. However, farmed oysters get shipped further and further inland all over the country, creating hundreds of thousands of new shells that wouldn't have been in the ecosystem naturally—that then need to be disposed of.

The practice of sustainable harvesting has more benefits than just providing a stable, renewable food source; intact oyster beds prevent erosion and keep water quality high. Oysters are very effective at cleaning water with large oysters filtering thirty to fifty gallons of water a day, making them extremely important to the health of local water systems. In many cases, this could save tons of money that would otherwise go toward efforts to clean polluted water, restore eroded shorelines, or mitigate farmland flooding. Oysters are not going to stop being shipped around the country, but by working to restore and preserve these oyster beds, attention is brought to the issue of mislabeling something as waste that has extraordinary value and potential to be reused.

Negev region of Israel—formerly part of the Byzantine Empire. Research into the trash left behind gives clues as to when the city fell, because "the end of trash collection—a crucial city service—would mark the end of high-level societal functioning in an urban environment."[11] It was only by looking at the remains of trash that archaeologists and historians were able to more accurately put together a timeline of when Elusa ceased to exist.

Ancient Landfills

New information about systems of trade and the economy of the Roman Empire has been found at Monte Testaccio, a 160-foot-tall hill in Rome's center. The mound is an ancient landfill, started possibly as early as the first century BCE. It's made up of an estimated twenty-five million shards of amphorae, a type of handled clay jug, that were discarded from shipping imports along the Tiber River. Imported goods like oil and wine made the huge jars difficult to reuse after the substances leeched into the porous clay, often going rancid in transit. These jugs were labeled with their place of origin, the weight of their contents, and to whom they were being delivered—evidence of a highly regulated and controlled trade system.

There was an intentional system behind how and where the fragments were placed on the mound to avoid landslides and maintain air circulation to keep the smell at bay, which allowed it to be a gathering site for community religious rituals and city festivities. Having a city's festivities take place at the city dump is a testament to its cleanliness. These safety and sanitary measures remain effective to this day. It's sturdy and dry enough that, two thousand years later, it now serves as a wine cellar carved into the side of the mound.[12]

Several hours south of Rome, a sophisticated sorting system was discovered, preserved in the ashes of Pompeii. We mostly think of Pompeiians eerily frozen in time at the moment the volcano killed them, but the ruins also show how those residents sectioned off different materials for different uses and reuses, such as construction materials for walls and floors. Soil samples illuminate the different materials dumped in certain sites, with organic materials producing richer soil than areas where tile or plaster were kept, giving a picture of how this society functioned. Much like a modern landfill, trash was relocated from the center of the city to the outskirts to be sorted before reuse. This removal system created the perfect conditions for archaeologists, as these sites are highly preserved by virtue of being ignored.[13]

Allison Emmerson, lead researcher of the Pompeii trash excavation, tells *The Guardian*, "The Pompeiians lived much closer to their garbage than most of us would find acceptable, not because the city lacked infrastructure and they didn't bother to manage trash but because their systems of urban management were organised around different principles."[14] When the mentality around trash is that it's a renewable resource, it becomes less of an eyesore and more a pile of opportunity. This way of living hasn't disappeared. For instance, I live out in the country, and there are many properties around me that are essentially self-contained junkyards, often overflowing with old cars, laundry machines, beat-up barns, and piles of wood. I see a version of my future that includes . . . *piles of opportunity*: endless possibilities to deconstruct and rebuild to keep myself entertained with never-ending junk projects.

What Ancient Trash Systems Tell Us

Trash is ignored in countless ways. It's seen as undesirable to have around or unimportant to think about, but because we overlook our trash, we miss the opportunity to see the creative, communal, and societal changes that are sometimes only visible in what we throw away. When we choose to not look at what we throw out, and we as individuals are not responsible for disposing of our waste or that of our neighbors, our attention is no longer turned toward what we discard, but rather solely on what we want to consume. As humans have developed larger and more complex cities, we have evolved to have not only better systems of shooing away unwanted materials, but also worse values around what is considered waste. As Emmerson states, "The countries that most effectively manage their waste have applied a version of the ancient model, prioritising commodification rather than simple removal."[15]

TRASHY TIDBIT

Animal Middens

HUMANS AREN'T THE ONLY ANIMALS TO CREATE MIDDENS

Octopuses, squirrels, rhinos, muskrats, earthworms, and other animals dump materials in designated locations.

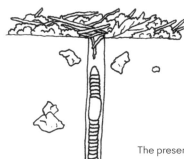

The presence of worm middens can indicate soil health—the more middens the better. Night crawlers can burrow as far as six vertical feet into the soil, building middens of plants and (their own) feces at the opening of their burrows, which serve as protection from the elements and as a stockpile of food. Farmers welcome night crawlers, as their presence improves soil health by composting, aerating, and encouraging beneficial microbes to reproduce in the soil.

6 FT.

Red squirrels live up to their hoarding stereotype, collecting upwards of 15,000 pine cones hidden beneath piles of scale dropped from previously eaten cones. The discarded scales act as insulation for the stockpile of winter pine cones to stay cold enough to not open (as they naturally would in spring) and drop their seeds. Red squirrels will keep their ever-growing middens for years.

Octopuses usually live a solitary life in dens, piling up the remnants of what they've eaten—bivalves, crustaceans, and gastropods—in front of the entrance of their den for protection and potentially territory signaling. Or maybe, like us, they don't want to bother taking out the trash any farther than they need to.

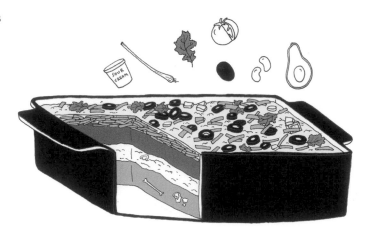

SEVEN-LAYER LANDFILL

Human evolution is documented in what we throw away. Currently, our top layers of garbage reveal that we have an excess amount of food in some locations, an abundant new material called "plastic," and variation in what and how much certain populations discard, and that our current layer is *much* thicker than any past layer over the same duration. There will always be lost material over time due to degradation, but even that can tell us a lot. What remains reveals so much about who was there and what they were doing. Our layer will long outlast anyone alive today and might outlive the entire planet's current species, depending on the way things go. And who knows what the ruins beneath a surprise volcano might reveal to future populations.

Chapter 2

SO, HOW DID WE GET HERE FROM POMPEII?

S o, what shifted in the cultural mentality to go from a philosophy of "use and repair" to one of "use and toss"? At its core, it's a shift in values: what we think is worth replacing and what we believe holds value to keep and repair. We have moved away from appreciating the materials and processes by which objects are made, and therefore have separated the objects from the value they hold.

Preindustrialization

Before production of goods was widely mechanized, back when all materials came from the earth (not labs), everything was sourced and created by skilled hands—and most hands were skilled in a craft of some kind. Blacksmiths made everyday cookware, farming equipment, and building materials, such as hammers and nails. Cobblers made and repaired shoes. Women, rich and poor, sewed clothing for themselves and their families, mending and reinventing garments until they were unusable as clothes. They were then turned into rags and used until they were turned into paper. Food was neither grown nor prepared in excess, and what was left was fed to livestock whose manure was used as fertilizer. Potters made jugs that could be repaired or used as dry storage if a crack formed. Furniture was carved and upholstered by hand, maybe even with fabric made from the wool of nearby sheep. Societies functioned because of and around skilled craftsmanship, specialty shops, and local economies.

Repair work in the preindustrial world evokes the Ship of Theseus, my favorite thought exercise. The ship of the mythical Greek king of Athens was repaired each year by Athenians to honor his legend and keep the boat intact. Failing boards were replaced over centuries, maintaining the structure of the boat while slowly replacing its entire physical body, plank by plank. This thought experiment is often used to demonstrate that object and material are not necessarily one and the same—the boat is still the boat, regardless of its material changing over time. When repair work is done to an object over time rather than replacing the object all at once in its entirety, it remains both physically and emotionally valuable. The repair work takes time but results in longevity, not disposability, of an object.

There are many ancient repair practices one can see around the world, albeit on a much smaller scale. One that stands out is the Japanese practice of *kintsugi*, which translates to "golden joinery." This process is a laborious and tedious act of preservation; when pottery breaks, the broken shards are joined together with lacquered golden seams made from the sap of *urushi* trees and gold powder. Cracks and flaws are celebrated and glimmer with precious metal—not only to mend the piece, but to add financial and emotional worth to it through practiced skill and highly valued resources. There is no verified origin of the technique, but it's believed to have begun in the sixteenth century and is part of the Japanese ethos of *wabi-sabi*, an embrace of imperfection.

In Europe, ancient Greeks commonly used lead to repair pottery due to its soft, malleable state when heated and its low cost and abundant availability. Repair work was more affordable than purchasing or making new pottery, so restoration was the most economical and reasonable way to sustainably keep vessels (and possibly get lead poisoning in the meantime).[1] And in ancient China, staples were often used to fasten together the pieces of broken plates—a process it's hard to imagine wouldn't shatter the plate even more.

The Decline of Craftsmanship

Reliance on craftsmanship and natural materials was practiced globally from the time of the earliest human civilizations up to the Industrial Revolution, when production shifted from skilled individual labor to large-scale machine manufacturing. This introduced a more efficient and robust

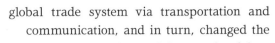

global trade system via transportation and communication, and in turn, changed the perception of value. It became less labor-intensive to make a product, and as human investment in its production went down, so did its overall value.

If you have created anything by hand that required a fair amount of labor and care—be it a pair of pants that fit perfectly because they were sewn by you, a chair you made from a tree in your yard, a pot you threw in a ceramics class, or even a simple knitted scarf—that object likely holds meaning and value. Those things feel more precious because you witnessed and partook in the process, even if just part of it, and most likely felt some ownership and pride over the finished object. If the chair gets a big scratch, you might eventually sand and refinish it; if the pants rip, you'll mend them, or, if the rip is major, turn them into shorts to prolong their life. We seek to repair things when we want them to last and know they're not easily replaceable without a lot of work. I have made a chair, sewn pants, knitted a scarf, and long ago made a clay pot—all of which can be tedious, labor-intensive, rewarding, and often frustrating. Doing these projects makes obvious the skill required to truly master a craft and helps us appreciate objects created directly by hands—especially our own. Craftsmanship is no longer a career held by a large percentage of the population; instead, "making and repairing things have become hobbies, perhaps not yet exceptional, but no longer typical."[2]

Instagram Craftsmanship

The COVID-19 pandemic spurred a sense of nostalgia for skills that most wealthy Gen Zers and millennials never had. This was called "cottagecore." When lockdown happened and people were stuck at home needing escapism, activities such as growing one's own food, sewing, embroidery, and baking bread became social media staples. These (mostly wealthy young white women) made a trend out of Eurocentric farm life from the past that was certainly much more difficult. A reprieve from the work/life hustle allowed those with financial security to slow down.

The pandemic made it clear that our work system in the United States is not healthy or equitable. A forty-hour workweek that focuses on productivity isn't sustainable, and we should all have the ability to slow down to focus on our lives rather than our work, but realistically, most people can't make that choice. The fetishization of labor, rural life, purity, and handicraft repair work by people who don't have to do those things for money highlights how disconnected our relationship to repair and workmanship is. It also ignores the class and racial disparities of these types of labor, excluding from the internet cottagecore airwaves many Black and brown people who have histories in rural craftsmanship, often rooted in colonization.

Assuming the aesthetic of an idyllic, utopian, flowery rural life of escapism ended up fueling capitalism, which happily commodified the cottagecore look in fashion and trendy decor. A cursory search into the world of buying cottagecore clothing leads you to sites offering $200 prairie dresses and housewares, which will at some point fall out of favor and create waste. Without acknowledgment of the realities of life for non-white farmers or craftsmen, past and present, or what rural life looks like outside of the English countryside, it can end up feeling like the well-intentioned rejection of capitalistic consumer culture does just the opposite.

I do not like baking bread—there, I said it! It takes five hundred hours and it's almost guaranteed to taste worse than bread from the bakery nearby. I'm not saying people should never bake their own bread, garden for fun, frolic in the grass, or sew for a hobby. These hobbies can be enriching, important, grounding, and fun, but it can feel as if the point of cottagecore's aesthetic is to exist online, which I'm pretty sure isn't why farmers farm or why repairs are made—out of economic necessity.

Our Backward Modern Economy

I took a turn into cottagecore for a moment there, not just to rant about it, but to show how detached we are from the realities of what it takes to produce and repair: to grow food or sew clothes in any way outside of a hobby, as well as how trendiness perpetuates capitalism and waste. The value of objects changes when we never see how they're made; from the millions of acres of cotton harvested to the metal mined from mountaintop removal, not having a hand in the process means we don't see the destruction required nor the value of labor put into creating almost everything. In most realms of production, machinery has replaced human labor, and with it, skilled craftsmanship. What's left of skilled labor such as cobbling, upholstery, and pottery is more expensive and harder to find, while skills like sewing or welding are undervalued and overworked. Our system is now backward; the skills required to repair are more costly and scarce than the skills required to manufacture new products.

In this flip-flopped economy, textiles stand out as a good example of devalued skilled labor. Almost no clothing is fully made by machines, as they aren't sophisticated enough to replicate human skill. We have technology that allows people to *live in outer space* on the International Space Station, but sewing every T-shirt in the world necessitates manual labor from expert hands. While some might find that to be a shortcoming of robots, it is a testament to the fact that everything requires human labor at some point in the process, and some human proficiencies cannot be replicated by machinery.

Cost over Quality

The battle of cost versus ethics is a major reason behind our trash problem. Your four-dollar T-shirt was sewn by someone, likely in a low-income country or by an undocumented worker in the United States who was paid essentially nothing. Workers in Los Angeles sweatshops have been documented being paid as low at $2.68 per hour,[3] and overseas, garment workers in India can be paid as little as fifteen cents an hour.[4] That four dollar shirt has worth because it was sewn by human hands and grown from the earth, but the valuation of the object does not reflect those resources because we've been told by the price that it's disposable. We don't see those skilled hands or the numerous stages of production leading up to and following the actual sewing of the shirt. When something is shelved in a stack of fifty items just like it, it becomes dangerously easily to replace when it wears out. And it doesn't help that celebrities like Justin Bieber are bragging about their ability to use and toss items: "[You] don't wear the same pair of underwear twice," he said, regarding his boxes upon boxes of free Calvin Kleins.[5]

The allure of instant gratification dominates our drive to acquire new objects; we often don't consider things like long-term functionality or satisfaction and, in turn, the waste it will produce. That four-dollar shirt is likely going to have a shorter shelf life than a thirty-dollar shirt made from more responsible materials, sewn by (more) fairly paid workers, and constructed for durability, and that (possibly) comes with a lifetime warranty for repair by the manufacturers. For those that can afford the luxury of choice in this way (and that number is small on a global scale), the pricey item will likely outlast the cheaper option, but the cheap item has the potential to satisfy the desire for newness without commitment to a style choice for the future. When you acquire something durable, there is a mental calculation you must make to decide if it will remain valuable to you. Paying a higher up-front cost elicits fear of potential obsolescence down the line. The entire notion of a short-lived fashion trend feeds this mentality. Wanting to avoid buyer's remorse encourages waste and impedes decisions that are often more ethically responsible. However, investing more in a purchase will likely create a higher appreciation of the item over a longer term.

The luxury of choice is inaccessible to many who, for affordability's sake, must opt for cheaper goods that will break down more quickly. In the end, this becomes more costly to those who can't afford the higher-priced, well-made goods. This goes for T-shirts, cars, furniture, and many other objects we use on a day-to-day basis. I have one very nice pair of shoes that I bought new several years ago. I have worn them well over a thousand miles, and they are only now starting to wear a bit. They're the best pair of shoes I've ever owned, likely the most expensive, and, unfortunately, one of the least fashionable. While they weren't cute at the time and still aren't now (waterproof hiking shoes aren't known for their fashion), my privileged decision to purchase something expensive once has saved me a lot of money in the long term and let me cut down on throwing out three mediocre pairs of shoes in the same span of time. This is a common pattern of wealth and poverty; being poor is more expensive much of the time.

The Allure of Newness

In the grand scheme of waste production there are of course many exceptions; cost and quality are certainly not always correlated, and usually cost is inflated across the board to make items seem more valuable for the sake of name recognition, status, or brand allegiance—even if their quality is the same. Consumerism demands that objects be seen as temporarily high value but worthless in the long term to maintain high levels of production and profit. Companies are not quick to reveal the ethics of their production practices unless their pitch and public appeal are based on ethical sourcing or production. Brands such as Everlane and Patagonia attract audiences because of their more sustainable sourcing and production; their pitches are largely based on *how* their products are made, rather than how affordable or accessible they are. If a company's appeal is not specifically based on its manufacturing, most will opt out of revealing any of the behind-the-scenes details: where materials are sourced, who is working to produce them, and how the manufacturers handle their waste.

Even if a product's quality isn't the best and we're buying based on a trend or brand, we still tend to emotionally invest in objects that we have invested more money into. Two extremely ubiquitous principles of upholding this pattern of constant product turnover are *planned obsolescence* and *dynamic obsolescence* (also known as *psychological obsolescence*). They are distinct but have overlap that you'll very much recognize if you are a consumer, which almost all of us are.

Planned Obsolescence

Planned obsolescence is the practice of intentionally designing and manufacturing products to have a short life span by virtue of poor quality, prohibitively expensive repair, or quickly outdated incompatible software and/or hardware. This strategy ensures that, through brand allegiance and necessity, products must be replaced, with people generally buying replacements from the companies who made the faulty products. Businesses figured out that from an economic standpoint, consumers buying one product that would last for years or decades would render their companies financially static, if not in decline. If every household has an oven that lasts fifty years, not many ovens are going to be sold. But if warranties are brief, parts are faulty, repair is expensive, and new models are sleeker, the number of ovens being bought, and thus discarded, drastically increases. The discarding of old objects is usually an afterthought to the acquisition of the new, exciting item replacing it.

One of the classic examples of planned obsolescence can be seen hanging in the Livermore, California, fire department. The Centennial Light is a much-beloved Guinness World Record holder: a light bulb that's been continuously lit since 1901. There is a webcam trained on the bulb that streams twenty-four hours a day, and the bulb has outlived three webcams over the course of its streaming history.[6] People can't get enough of this light bulb, and those who are up on their planned obsolescence history know why: light bulbs were the canary in the coal mine of a new wave of consumer capitalism.

In 1925, the Phoebus cartel (pretentiously named after the Greek god of light) was formed by a number of incandescent light bulb manufacturers, including but not limited to OSRAM, General Electric, Associated Electrical Industries, and Philips. This group was one of the formative forces behind the widespread development of intentional planned obsolescence. If every major light bulb company agreed to make a worse product, they all benefited—even if they remained in competition with one another. Before this alliance, incandescent bulbs could last upward of 2,500 hours, but as more homes began to have access to electricity, the Phoebus cartel needed the life span of their bulbs to be significantly reduced to make this new production model work.

For several years, these companies put their engineers to work on making their products *worse* (to burn out after one thousand hours), which in turn created a deluge of light bulbs being thrown away. This effort was a huge success; bulbs performed poorly across the board. Strict monitoring was put in place, and companies that went over the thousand-hour rule were heftily fined. The incentive to swindle customers was strong enough that they held one another accountable, despite being economic rivals. Once World War II began, the global alliance disbanded, but the impact of the idea they established is seen in products across consumer categories today.[7] Though at least in this one instance, we're seeing a positive development. Almost one hundred years after the cartel pioneered planned obsolescence, production of incandescent bulbs is now banned in dozens of countries around the world with most others implementing strategies to phase them out in favor of much longer-lasting LED bulbs designed to reduce waste.

I know we've already covered a sprawling time span here, from ancient trash systems to the trash systems of today. This is all to point out a particular irony. In the past (pre–Industrial Revolution) we lived sustainably out of necessity. Our systems and our "stuff" were made to last, to be reused, or to serve a new purpose after being used in the first place. And now that we've moved away from those systems, embracing a more destructive, waste-acceptable lifestyle, we're in an environmentally dire situation that is forcing us to revert back to the practices from the past—increasing repairability, efficiency, and recyclability.

Psychological Obsolescence

Around the same time that nefarious plans were being made in the light bulb arena, psychological obsolescence came onto the scene in the automobile industry. *Psychological obsolescence* is why you might have twelve different styles of cell phone charger from the past six years. The chargers aren't broken, but they don't fit into the ports of the newest yearly model of phone, computer, or headphones. In this design strategy, marketing plays a significant role; the desire to be trendy, current, or the most technologically advanced drives replacement consumption even when existing items are still in perfectly good condition or new ones are barely different from the previous models. Marketers and designers realized "they could impose obsolescence by desire, making customers want new products so intensely that they couldn't wait to trash their current ones."[8]

In the 1920s, while General Electric tinkered to make light bulbs worse, General Motors worked to make cars more exciting, sexier, and trendier—a product people would voluntarily choose to upgrade rather than purchase as a replacement for something broken. It was a trick of desire, not need. Up until that point, Ford was the leader in automobiles, manufacturing the car we think of as the ubiquitous old-timey car: the black Model T. Over the course of nineteen years, Ford manufactured fifteen million Model Ts, keeping them affordable but aesthetically unchanged—an expected dependability that would soon backfire during a time of big cultural shifts in consumer behavior.[9] In 1925, a new Model T was $260, which is about $4,500 in 2023—a price point *far* lower than today's vehicles.[10]

At the time, Americans were starting to desire trendiness over practicality, and Ford's competitor, General Motors (GM), capitalized and accelerated this cultural shift by introducing cars as a status symbol. Owning a car in and of itself showed wealth, but GM wanted to create status within status: a variety of makes, models, and colors. They produced lines of cars that went from economy to luxury, with different features and styles in each line. Unlike the Model T, these cars allowed drivers to put their socioeconomic status on display, one could say. These luxury cars didn't function better than those in the economy line, but aesthetics often mean more in the scheme of physiological obsolescence.

Up until that point, durable color paint technology beyond black and white hadn't been developed, but GM partnered with DuPont to engineer colored car paint, which had a surprisingly huge impact on car desirability. GM realized that cars geared toward women—fashionable and more feminine—would satisfy the craving for newness and trendiness, with color options playing a large part of that appeal.[11]

In order for this marketing and manufacturing manipulation to work, new styles had to constantly change, relying on the idea that "psychological obsolescence fashion depends on disappointment"—as soon as you have the shiny new car, you're anticipating the next model.[12]

Though General Motors wasn't the only company using this model of sales at the time, it was most certainly the catalyst that led to such staggering levels of consumerism, and in turn, waste production. People learned to look toward objects to feel pride and success: an achievement of wealth or style as represented by the stuff one owns.

In the theme of returning to past practices that were intentionally squashed for the sake of obsolescence, GM's electric car is a great example. As the instigator of much of planned obsolescence's ubiquity, GM has come full circle on their own product: the electric car. In the early 1990s, GM invented the EV1, the first fully electric car on the market, as part of California's zero emissions mandate that required major auto manufacturers to develop electric vehicles. Fewer than 1,500 cars were produced, and despite consumer interest, under industry and government pressure along with poor profit margins, GM shut down the manufacturing of the vehicles and subsequently smashed all of them save a few that now live in museums. A full twenty years later, GM is circling back to electric vehicles as demand grows with worsening climate change. GM squandered the opportunity to be ahead of the game when it came to technological development in the realm of sustainability. But sustainability and profits, as we've seen, have been viewed as a threat to one another, contributing to mass waste when profits win out.

Apple: The Wasteful King of Planned and Psychological Obsolescence

Consumers have a somewhat limited choice in the matter of planned obsolescence, and therefore contribute to the growing problem of e-waste production. We depend on phones, computers, and cars in daily life, and when they're broken, we aren't given many options to replace them with better technology—remember, many newly manufactured goods are designed to break. However, we are able to choose who we buy from and why, and that's where dynamic obsolescence comes in. I got a cell phone in 2001 to call my parents to pick me up: the Nokia 3310 brick phone—the one with Snake and T9 texting that lit up on the sides—and it was nearly indestructible. Lovingly nicknamed "The Terminator" or "The Unbreakable" and

the subject of an internet cult following, this phone is a good example of a durable, reliable product that was overshadowed by a more stylish, novel product: the smartphone, specifically Apple's iPhone, which was first released in 2007. This new product launched a dramatic increase in electronic waste being a household norm. It's estimated that the average cell phone now has a life span of 2.5 years. As of 2017, a guy named Dave Mitchell has been using a Nokia 3310 for 17 years, stating, "If it ain't broke, it's probably a Nokia 3310."[13]

It's sometimes hard to know whether the poor performance of technological advance-

ment is intentional or, by virtue of new materials and design, meant to meet consumer desire. As we crave sleeker designs, we seem to get more break-able products, which makes sense in some basic ways. A phone encased in a chunk of hefty plastic will most likely be sturdier than a thin phone with a fully glass front and tech that relies on touch screens. The latter will prob-ably encounter malfunctions more quickly than phones with analog buttons, but it's hard to accept that the design of something that is only becoming more ubiquitous sacrifices longevity for the sake of sexiness. Our evolving desires for new product features are determined by the products we're given (and subsequently, what we throw away). We want a bigger phone once we see there's one available, then they get bigger and bigger and suddenly they don't fit in any pockets. Then that's too big, so now they're getting smaller and smaller and essentially back to the size they were in 2007. And in the course of that size roller coaster, billions of dollars were made by showing people what they didn't know they wanted until they saw it, which is a cru-cial part of psychological obsolescence: constant innovation drives constant consumption, even when the innovations do not reflect shifting needs. If we're back to the original-size phone, we probably never needed a bigger one.

Apple is notorious for being a perpetrator of planned obsolescence that is deeply rooted in brand allegiance and style. I'm one of the billions who, begrudgingly, can't quit Apple. Do you have fifty different cables jumbled in some untouched, shamefully chaotic drawer that don't seem to fit any-thing you currently own, or only sometimes seem to work? That's why. Their branding is sleek, beautiful, and futuristic, and their users, myself included, rarely stray from the brand, despite their products being worse in many ways: not water-resistant, prone to decreased functionality by software updates meant to improve them, prohibitively expensive, or impossi-ble to repair. Apple is constantly churning out new products that are, in re-ality, only *marginally* different than the models before. Many of the cords in my cord chaos are generic versions that are more readily available and cheaper, from the gas station or store when I am on a road trip or have left my charger somewhere. Because branded products are often much more expensive for the same thing, it ends up creating a market flooded with cheaply made versions that break and are eventually thrown out after be-ing taped together as many times as possible.

My current iPhone is, according to Apple, supposed to be water-resistant if submerged in up to one meter of water for up to thirty minutes. About a month ago, I dropped it into a cup of water for no more than thirty seconds and the speaker broke. I had an iPhone bend over time from just being in my pocket (part of what has become known as "Bendgate"). I've had screens shatter from a one-foot drop. I will admit there is a theme of being clumsy and close to water, *however*, these expensive products that are labeled as water-resistant should be expected to have more durability, or at the very least live up to their claims of durability. The Deepsea Challenger submarine has 3D cameras and robotic arms that function at a depth of 35,787 feet into the ocean's Mariana Trench . . . Apple can make a phone that is waterproof.

My parents had a teal bubble Mac when I was a kid in the nineties, and since then, despite having choices, I keep choosing Apple even though I hate their lack of longevity; their software updates generally decrease the processing speed; and it's almost as expensive to fix the phone as it is to buy a new one. Midway through writing this book, my eight-year-old laptop broke and I had to replace it. Clearly quality hardware is not the reason smartphones outpaced durable phones like the beloved Nokia. Our technological desires have advanced and changed the way we navigate daily life. We evolve alongside our objects, one influencing the other. The hard reality of these models of obsolescence is that consumers are now complicit despite having options that are better made. I feel guilty about the phones I know need to be recycled or tossed—they didn't last long and elicit consumer guilt due to a poorly manufactured product. Whether because they broke or because their owners were drawn in by the allure of new models, in 2022 it was estimated that 5.3 billion cell phones were thrown away.[14] These are not all Apple—many tech companies employ tactics of obsolescence to outdate their own products.

When I looked up the average life span of a MacBook, I learned that my 2015 computer was categorized as "vintage" by Apple in 2021. According to Apple's website, their products "are considered vintage when Apple stopped distributing them for sale more than 5 and less than 7 years ago." The general rule of what qualifies as "vintage" is that it refers to items more than twenty years old (but less than one hundred years, the age at which items are categorized as antiques). By self-proclaiming their own products as vintage after just five to seven years and refusing to service those products after that window of time, Apple recognizes and openly capitalizes on their planned obsolescence. I am regularly tempted to return to the way of the non-smartphone to remove myself from the grips of Apple and smartphone addiction, but I likely never will. I need to access the things I've become accustomed to relying on, so I will begrudgingly keep all my tech ten feet away from water and pray Apple comes through with their waterproof promise someday.

There is an increasing amount of dissent against companies like Apple, including class action lawsuits against their software and hardware, as well as government legislation in the EU moving to require all devices to transition to using only USB-C cables, a universal port, which will make it illegal for Apple or other companies to outfit their devices with proprietary chargers. My new computer has two USB-C charging ports and nothing I have is compatible, so while it's a great move forward to stabilize the ever-changing cord chaos, I have to, for hopefully the last time, get adapters or replacement cords for all my USB-A cables. The benefits to such a change are massive and wide reaching; switching to a universal charger will save an estimated twelve thousand tons of e-waste per year.[15]

The universal cable legislation is a great example of a very straightforward, specific change by corporations (even if it's by force) that can have a huge impact on consumer behavior and its subsequent waste generation. Before this law, we would choose to buy new devices but not choose what their required cable shapes were, forcing us to toss out the old and buy new with every upgrade. Tossing out the old usually means stuffing them in the cord graveyard of your closet with the random router cables, broken chargers, and mysterious cords that came from who knows what, only to all be tossed years later during a move. By giving consumers only one option, it eliminates the choice altogether when it comes to needing to cycle out outdated tech gear. From product to product, we can keep the same charger no matter if the product it came with has broken or become obsolete.

The Road to Responsibility

Light bulbs, cars, phones—they're all things we rely on for daily life. We rely on the decisions made by companies as to how well our products work and how much we have to pay to maintain or replace them. Value has been determined on a capitalistic cultural level—a combination of our desires for

TRASHY TIDBIT

An Inventory of Oscar the Grouch's Trash Can
(According to Muppet Wiki[16])

Television set, piano, bowling alley (adjacent to his billiards room), billiards room, living room, chiffonier, library, foyer, greenhouse, kitchen, fountain, credenza, shower, velvet sofa, bearskin rug, Aretha Franklin records, crystal chandelier, washer–dryer–folder–putter–awayer, den, basement, rumpus room, oven, number collection, train, fireplace with mantelpiece, dining room (located in a swamp), submarine (yellow-colored), floodlights, tennis court, squash court, china cabinet, bedroom, pickle bush, ballroom (led by a lobby in his west wing), sauna (filled with a dead lobster collection), game room, statue of Uncle Ugly, transit (bus) system, stereo set, dump truck, lighting, hope chest, 1937 Oldsmobile (no wheels), conservatory, trampoline, framed picture of Benedict Arnold, family, suit of armor, pool, trash compactor, Ping-Pong table, vacuum, cabana, zoo, mudroom, elevator, chicken coop, alligator lounge, grime room, drawing room, family room, knickknack shelf, darkroom, furnace, twenty-step spiral staircase (nineteen good, one broken), pigpen, lily pond, hot tub, toy boat, The Queen Muddy.

novelty and status, coupled with a lack of access to more durable products like the Nokia brick phone. Consumer choice sometimes matters, but choice is limited because of corporate bottom lines: how to make the most money, which often translates to making the most stuff last the shortest amount of time.

We waste in part because we've learned that things only have temporary value; once they no longer hold our interest, they're trash. We as consumers have the option to resist the temptation of unnecessary upgrades, repair over toss, and buy used when possible. Corporations should have the responsibility to cease their habits of intentionally forcing those upgrades and setting up roadblocks for repair. Having reusable items around creates a system that requires less consumption while not forcing people to change their lifestyles all at once. In order to make these cultural changes, producers need incentives to design and manufacture goods that are better for the planet—because in the current system, they economically benefit from creating throwaway products with no responsibility to make different decisions. There is rarely altruism in corporate strategy.

One of the most effective ways of tackling this issue is Extended Producer Responsibility (EPR). This strategy is designed to transfer the burden of disposing of postconsumer products back onto the companies that manufacture them, rather than passing on those responsibilities to municipalities and taxpayers. EPR policies encourage design updates to reduce waste management costs. The more recyclable a product is, the cheaper it is for the company to use recycled materials from their own products rather than virgin materials. The goal is to increase the circularity of a product's life and reduce the environmental toll of manufacturing. Big corporations operate around money and have a vested interest in cutting costs while still selling as much as possible. Recent efforts to greenwash products like Coca-Cola's biodegradable bottles and ad campaigns by H&M and ExxonMobil reflect an effort to look good but not actually do better. Starbucks attempted to push a redesigned "strawless lid" when straws became the environmental enemy, but it was quickly pointed out that their new, "greener" lid actually used more plastic than the one it was replacing (including the straw). These are publicity stunts—not real efforts.[*]

There are a couple of ways EPR policies can look—some are voluntary and some are government mandated. Advanced recycling fees (ARFs) are collected from producers and put toward fees associated with disposing of their products. "Take-back" policies require companies (manufacturers or

---

[*] As of 2023, the United States is the only member of the Organization for Economic Co-operation and Development (OECD) that does not have national EPR legislation.

retailers) to take back products once they're done being used. This responsibility creates financial incentives to design products that have reusable parts, fewer materials, and materials that contain less toxins, and that require fewer brand-new parts. If they have to pay for the waste management costs associated with their own products, they'll be more creative, and conscious of smart, environmentally sound designs that move away from planned obsolescence.

In a closed-loop economy, goods are produced and disposed of in a self-sustaining way. Products are designed to be reused, repaired, and/or upcycled into high-quality, non-degraded materials that enter back into the circular manufacturing model. Material extraction and product manufacturing generate far more greenhouse gases than the actual use of those goods. When done in a smart, efficient way, circular design has the potential to be cost-effective for production and disposal, create jobs, encourage innovation, and reduce waste during material extraction and manufacturing.[17]

So much of what will be discussed throughout the rest of this book are the effects of a linear cradle-to-grave use model versus a circular cradle-to-cradle one. We're having a "come to Jesus" moment that things need to change. We're struggling to break the normalcy of easy, frequent disposal, and move toward the types of circular economies that existed in the preindustrial world—away from our extremely linear mode of use and disposal. The change is not happening nearly as quickly or widely as it needs to, particularly in the United States, where corporate entities have much more power than they should and governments are slow to enact any sort of accountability upon them.

Moving toward cradle-to-cradle practices is what will alleviate our garbage crisis. We've strayed far from the art of stapled pottery, adult pants turning into children's pants, and sustainable farming shell mounds. We're harming ourselves and future generations with our current unsustainable lifestyles. Garbage kills the people that must live with it in their communities, and for those of us who have the privilege to pretend it doesn't exist, the less "out of sight, out of mind" we can make trash, the more people will understand the situation we're in. Buckle up for some depressing information that will make it all a little more clear.

Chapter 3

THE BEGINNING OF THE GARBAGE INDUSTRY

Despite how different the world is today than at any point in history, the basic principles of garbage disposal haven't changed. The methods evolved in their mechanization and scale to accommodate growing civilizations and newer technology, but dumping, burning, recycling, and reducing have always been the way to dispose of what we create—natural and synthetic materials alike. We pile up waste, whether that's human and animal poop in the streets, food scraps out the window, or in huge landfills today; we burn bodies, plant debris, and now all types of garbage in massive incinerators; we reuse broken pottery for vessels, recycle glass bottles, and reupholster furniture; and lastly, we have, with varied success, attempted "source reduction"—the very basic concept of using less. Over time and across geography, these practices developed in different ways and to varying degrees, from unregulated open-air dumps to an incinerator that has a ski slope built into it.

Around the world, garbage collection takes place in many different ways, from Beethoven-blaring trucks to pneumatic tubes to waste-picker collectives. The waste industry manifests differently largely depending on a country or region's landmass, political climate, and level of economic resources.

Taiwan is one of the leaders in modern waste prevention, reuse, and efficient collection, and it's certainly the undisputed leader in making trash collection fun, social, and effective . . . unless you're not a Beethoven fan. To follow the government-decreed "trash doesn't touch the ground" ethos, the small, densely populated country formerly known as "garbage island" cleaned up its act in a major way, creating a sparkling-clean, orderly, and conscientious society.[1] Taipei, the island's capital, was overrun in the 1990s by waste produced from a rapidly growing economy and increased individual spending power, expanding the daily municipal solid waste from 8,800 tons per day in 1979, to 21,900 tons in 1992.[2]

Landfills rapidly filled up the inadequate allotted space, and pollution got worse because of it. No incinerators or large-scale recycling systems existed. Because the island is small and dense, the options of where to put trash were limited, which is ultimately what led to creative problem solving. Increasingly angry about the government's lack of action to address the impending crisis, residents demanded the government's attention through protest and blockades against new and existing landfills. While a lack of space is challenging for myriad reasons, it forces one to see that waste doesn't disappear. It has to go somewhere, and when that somewhere is very close by, it's much harder to ignore. The United States doesn't face the same issue of space, which allows our ingrained sense of "out of sight, out of mind" to persist.

Through a mix of corporate responsibility and consumer participation, Taiwan created a highly successful system that results in very little waste. On the corporate side, there is an extended producer responsibility act that requires companies who manufacture or import goods (waste generators) to contribute funding to recycling programs.[3] While this by no means creates a circular system, it alleviates pressure from consumers to be the do-gooders while those upstream face no consequences. How we move away from excess, both in production and waste, will ultimately need to be a top-down change. Altruism and capitalism are almost always mutually exclusive, which we can see throughout history. Without penalty or accountability from those in power, there is no incentive to change.

On the consumer side, Taipei's process is highly participatory if you want your garbage to be out of sight. Twice a day, a large yellow truck followed by a little white one drives through the streets loudly playing Beethoven's "Für Elise" (or sometimes "A Maiden's Prayer" by Tekla Bądarzewska-Baranowska) to summon residents to bring out their trash.

Residents pay for city-issued garbage bags—an incentive to save money by producing less waste. To deter people from dumping household trash outside the home, thus avoiding the fee for trash pickup, almost all public garbage cans were removed, leaving just 1,700 cans in Taipei—one for every 1,500 residents.[4] By contrast, New York City has 23,000 cans, which works out to be about one can for every 380 people.[5]

Food waste is not wasted, but rather, continues to provide nutrients as livestock feed. Taipei's food recycling rate is about two-thirds of their total waste production, and sustains a huge pig population—a full circle food loop. This system is economically and environmentally more sustainable. Household food scraps are picked up from homes on the same nightly route, and are sold at a substantial discount to pig farmers; reusing these scraps saves land since the need for livestock feed farms is eliminated; and it reduces the energy-consumptive process of incinerating wet materials, which is Taiwan's main disposal method.[6]

Horse Shit

In 1900, horse-drawn carriages and buses dominated the streets of London, amounting to fifty thousand horses at work in the city.[7] In an 1894 article in *The Times*, it was predicted that "in 50 years, every street in London will be buried under nine feet of manure." Urban planners were faced with a difficult problem: How do you clean an excessive amount of horse shit when the method of cleaning it requires horse-drawn carriages? Deemed the Great Horse Manure Crisis of 1894, and also known lovingly as the Parable of Horse Shit, it's a good summary of the plight facing major cities around that time.

In the late 1800s to early 1900s, the problem of garbage production was beginning to exceed the capabilities of individuals to handle the garbage they were generating. Ash and manure were driving forces of organized trash hauling across big cities in the United States. New York was being buried in manure from the more than 150,000 horses living in the city, dropping a hundred thousand tons of manure and ten million gallons of urine per year in its streets.[8] Most major cities struggled with massive amounts of horse waste accumulating in the streets, as well as horses who died and were left there. Once bought by farmers to use as fertilizer, the supply of manure outgrew demand, leaving a huge, putrid mess that required 1,200 street sweepers to respond to New York City's manure and dead horse problem.[9]

This problem got the attention of New York's government to mobilize the first citywide collection service. George E. Waring Jr., a drainage engineer for the construction of Central Park, was hired to establish a sanitation system. Before coming to New York, Waring designed a sewage system in Memphis, Tennessee, that would separate stormwater runoff from sewage—an innovative approach that was adopted around the country. While he did not have experience in sanitation systems, he got to work putting together crews, known as "Waring's White Wings" for their all-white outfits, to clean up the waste. The white outfits were intended to represent professionalism and uniformity amongst workers.

The New York City Department of Street Cleaning quickly transformed the sanitary conditions of the city, reducing disease and increasing functionality. In 1894, Waring became commissioner of the department and implemented municipal services of street sweeping, recycling, and collection. The Street Cleaning Department would eventually change its name to the Department of Sanitation and become the largest sanitation department in the world.

In conjunction with the implementation of the Street Cleaning Department, soon after the invention of the automobile in 1908, horses were fully replaced as the primary mode of transport, solving the horse shit problem and unknowingly unleashing a whole other problem of impending pollution.

And Other Nineteenth-Century Problems

Outside the home, horses were defecating on city streets, while inside the home, coal was being burned for heat, creating an enormous amount of ash. You may have noticed that on modern municipal trash bins there's a message stating "no hot ashes." That's why. I always wondered how many people in 2024 could possibly put (or be tempted to put) hot ashes in their plastic municipal trash bin. In 1899, "no hot ashes" was a more relevant instruction, as household coal ash comprised 84 percent of total municipal solid waste (MSW)—a shocking metric both for physical volume of ash being produced, and the different distribution of what comprised MSW at the time.[10] Today, according to the EPA, that percentage would be equivalent to rubber, food, yard trimmings, plastics, metals, glass, and paper combined.[11] The Corona Ash Dump in Flushing, Queens, became the central location of NYC's ash collection in 1917, described as the "valley of ashes" in F. Scott Fitzgerald's *The Great Gatsby,* and where I went to see the 2023 Westminster dog agility competition.

In Chicago, the sanitation issues in the city were so abysmal that a group of women founded Hull House in 1889: a center that offered social services and housing to poor residents in the Nineteenth Ward.[12] Appalled by the city's filthy conditions—overflowing sewage, garbage, and high levels of disease and mortality—Hull House's cofounder, Jane Addams, led the campaign to improve residents' quality of life by addressing the putrid environment surrounding them. Known as the "garbage ladies," Addams, activist Florence Kelley, epidemiologist Dr. Alice Hamilton, teacher Mary McDowell, and the Hull House Women's Club collected data about the city's conditions, studied sanitation-related diseases, and generated information used to pass legislation and regulations—forming Chicago's first commission on waste.

Modern Waste

Luckily, dead horses, horse shit, and ash ceased to be an issue on the streets of most major cities in the early 1900s when home heating shifted away from coal, and cars replaced urban-dwelling horses. What took its place was *stuff*—man-made, machine-produced *stuff.* With an increasing amount and variety of stuff, the definition of what waste actually is has become more nebulous. The makeup of many of the things we use every day is more complicated than ever. Fabric is no longer purely plant fibers, but a hybrid of natural fibers and plastics; electronics are made of dozens or hundreds of materials; and plastics are a diverse group of relatively new substances. Sometimes materials are mixed for the sake of improving a product's performance, and sometimes it's done to lower costs (while also lowering quality).

The sheer volume of global waste production is higher than ever in history, at 2.2 billion tons annually. By "global waste production," I mean what we *do* throw away, not what we *should* throw away, which would be

TRASHY TIDBIT
Danbury Trashers

"We want to be the Evil Empire of the UHL. We want that bad-boy image."

These are the profound sentiments of AJ Galante, the seventeen-year-old president and general manager of the short-lived minor league hockey team the Danbury Trashers, represented by their mascot, Scrappy the garbage can.

Gifted the team by his father, Jim Galante, who created the franchise on April Fools' Day in 2004, AJ was in charge of recruiting players to make a team who deemed themselves the "bad boys of hockey." They were surprisingly good and unsurprisingly brutal on the rink, inciting violence and using tricks like greasing up their uniforms to avoid being grabbed by their opponents.[13]

The Trashers were not strictly an extremely expensive hobby, but a means of laundering enormous sums of money by adding players and their wives to the payroll of Galante's garbage service companies. At the time, the mafia was heavily involved in running the trash-hauling services of New York, Connecticut, and New Jersey, and Jim was the head of Automated Waste Disposal, a trash-hauling business in Connecticut.[14] His success in the waste-hauling business amounted to $100 million—a sum reflecting his efforts to push out all competition and monopolize the regional garbage industry.

Galante was under FBI investigation at the time for his involvement in mob-related crimes within the business of trash hauling, resulting in seventy-two charges and eighty-seven months of prison time.[15] With financial payoffs to Matty the Horse, a notorious figure in the Genovese crime family, he controlled a lucrative and competitive market, eventually owning twenty-five trash companies by the time of his arrest.[16]

Within three days of Galante's arrest, the Trashers shut down, ending the team's short-lived but memorable run.

a much smaller amount. For example, food scraps and wasted food—much of it is perfectly fine to eat, such as yogurt one day past its expiration date that grocery stores toss, or leftovers you didn't feel like eating the next day. Even though it shouldn't be waste, we treat it as such. With that in mind, the amount of waste we produce is staggering, and the breakdown of what makes up the bulk of that waste is both bleak and hopeful.

The Top of the Trash Heap

According to the EPA, in 2018, the top four municipal solid waste materials were paper products, food, yard trimmings, and plastics. The bleak part of this breakdown is the plastic. Abundant, ever increasing, and nondegradable, plastics prove extremely challenging to handle in a sustainable and responsible way; more on that soon. Hopeful possibilities exist in the realms of food waste and yard trimmings.

There are a couple of ways to categorize waste: what it is or where it's from. What it is might be plastic, wood, or metal, but where it's from refers to its origin, not solely its material. Construction, mining, wastewater treatment, and other sectors outside of the home or retail all generate waste, but are generally not included in tallies of MSW, which skews the overall waste generation dataset. Even within the picture of MSW there's murkiness. Would the carpet I ripped up and dropped at the dump be construction or household waste? How about batteries that end up in the landfill but are technically e-waste? Food waste that's composted or put down the garbage disposal but doesn't make it to the landfill? Mining waste that's mixed into a slurry rather than in solid weight or volume?

There is a measuring problem in the waste industry propelled by a lack of accountability enforcement across industries. A commonly referenced figure is that MSW accounts for only 3 percent of the total waste generated in North America—the other 97 percent being industry and agriculture. Globally, poor or nonexistent waste generation record keeping across all fields makes big-picture assessment challenging, and while it would be helpful to understand the ins and outs of waste, the chances of ever gaining an accurate picture are slim to none. The amount of labor required to decide what counts as trash and globally agree upon those categories, then keep accurate track of it, would be an unimaginable undertaking.

Garbage, at this point, is so omnipresent and nebulous that the numbers we can go by are not the full picture by any means, but it's safe to assume no one is overreporting. Most likely there are many more Olympic swimming pools' worth of rubbish than we're being told about.

TRASHY TIDBIT

New York City's Trash, Dead Horses, Rats, and the Mafia

New York City has long been hailed as the capital of the world, being home to the most wealth, diversity, cultural influence, and international diplomacy of any city on the planet. It lays claim to the most languages spoken (eight hundred), most millionaires in one city (more than 345,000), two million rats (Chicago still has more), the headquarters of the United Nations, and the world's largest sanitation department.

One of the most notable qualities of New York City that comes to mind, for some before they think of Broadway or the Empire State Building, is mountains of putrid garbage bags piled up on the sidewalks. It can appear that there is no rhyme or reason to where and when the bags are out, but there is a system. As we learned earlier, what began as a need to deal with massive amounts of horse manure and food scraps has evolved into a huge infrastructure of constant collection, plowing, and sweeping.

Kate Ascher explains the scale of the New York City garbage system in her book The Works: Anatomy of a City:[17]

> In order to function as smoothly as possible, a fleet of six thousand vehicles pick up garbage from 6,500 miles of streets and twenty-five thousand trash cans throughout the city, going through one hundred thousand tires per year. Their repair shop is 6 stories high, making it the largest non-military repair shop in the country. The ten thousand sanitation department employees are constantly working to clear the city's annual fourteen million tons of trash. Meanwhile underground there are 1.3 billion gallons of sewage per day traveling through 1,600 miles of sewer lines to end up at 14 wastewater treatment plants. The treated sludge is taken away on barges that transport three 300,000 cubic yards of sludge each day. Half of the sludge is made into fertilizer (mostly sent to orange groves in Florida).

New York has been in a constant battle with its sanitation since it grew into one of the densest cities in the world. Despite having the biggest fleet of sanita-

NYC'S (FAILED) ATTEMPTS TO ERADICATE ITS RATS

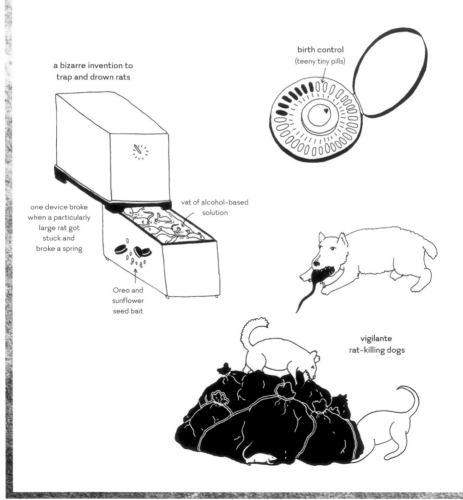

a bizarre invention to trap and drown rats

birth control
(teeny tiny pills)

one device broke when a particularly large rat got stuck and broke a spring

vat of alcohol–based solution

Oreo and sunflower seed bait

vigilante rat-killing dogs

tion workers, it lags behind big cities like Chicago, Singapore, and Amsterdam, who have developed efficient, clean systems of trash removal. In 2022, in an effort to clean up the subway system's garbage as well as the city streets, and to reduce the rat feast, NYC implemented fines for people who put their trash on the curb before 8:00 p.m., hired a "rat czar," and introduced the VakTrak (a train that vacuums the subway tracks), and is now hoping to containerize the city's residential garbage to further improve overall sanitation. All these initiatives have been met with complaints that go hand in hand with New York's need to, understandably, hate every mayor they have.

In 2017, Taylor Swift was fined $3,010 for thirty-two violations given by the New York sanitation department for failing to keep a clean sidewalk in front of her apartment.

A very small selection of
THE NEW YORK DEPARTMENT OF SANITATION'S
"TREASURES IN THE TRASH" MUSEUM.

In 1981, sanitation worker Nelson Molina Sr.
began collecting discarded items he found on the job.
Over the course of his thirty-four-year career,
Molina collected over 40,000 items, which
are carefully sorted by subject.

They're currently housed in East Harlem's
Manhattan 11 Sanitation Station.

Mayor Eric Adams is obsessed with the rat problem and has a desire for vengeance, creating the position of rat czar in his war against the rodents. He hired Kathleen Corradi, who he believes has "the drive, determination, and killer instinct needed to fight the real enemy: New York City's relentless rat population." Perhaps this is a personal vendetta, as in 2023, Adams himself was fined $300 for having rodents in his Bedford-Stuyvesant rental property.

Trash bags on the street are a staple of the city—as recognizable as the original "I Heart NY" logo (not the new one)—but they might begin to disappear as more trash bins emerge in their place. If you've spent time driving in New York City, you are familiar with the competitive game of parking Tetris: double-parking until someone moves, driving around the block to take your spot back when the street sweeper passes through, getting yelled at and having to check multiple complicated parking signs all on one post. . . .

Containerization might make that game even harder with containers potentially taking over 150,000 parking spots—also overhauling the truck fleet and its route.[18] As of May 2023, the containerization law was proposed specifically for food establishments in an effort to curb the rats feasting on "putrescible solid waste." Restaurant owners are worried that having private trash bins that are publicly accessible will be an invitation for others to dump trash, as well as block sidewalk areas in front of their businesses that could otherwise be used for seating.

When Trash Is Criminal

The people that are responsible for dealing with this big mess are the sanitation workers: a highly sought-after job with over sixty-eight thousand people applying for just five hundred positions in 2015.[19] If you're working directly for the city, it's a well-paid, stable job with unionized employees receiving benefits and pensions.

Since the NYC Department of Sanitation's (DSNY) establishment in 1881, all garbage was collected by the city (for businesses and residential buildings alike), but in 1957 when the city separated the responsibility of municipal and commercial waste collection, the commercial sector was left to private haulers: namely the Five Families of the Italian mafia—Genovese, Gambino, Lucchese, Colombo, and Bonanno—who split up the boroughs as well as part of New Jersey and Connecticut. Private businesses were suddenly responsible for contracting their own waste hauling services, and with a monopoly on each of the areas throughout the city, the mafia secured an enormous sector of legitimate business. Filled with intimidation, money laundering, and violence, the mob was unofficially in charge of the city's private waste-hauling industry until 1991 when the city cracked down on the mafia's stronghold. Across the Atlantic, in Naples, Italy, the organized crime group the Camorra ruled the waste-hauling industry for decades, including collection, illegal toxic waste dumping, and operating the landfills themselves.

The divide between privatized garbage collection and the municipal system is vast when it comes to treatment of workers, and the current state of privatized worker treatment seems like a small step up from the mafia. Private companies are notorious for having exploitative worker conditions with commonly occurring fatalities, serious injuries, low pay, and extreme overtime demands. The state of commercial trash hauling is chaotic at best and has little oversight or enforcement to improve a dangerous, environmentally damaging, and inefficient system. One small company with a limited number of trucks can wind up serving every borough, stretching its employees thin with nightly shifts that can last sometimes fifteen hours. Unregulated operations have earned sanitation work the position of fifth-deadliest job in the United States, with the vast majority of incidents happening in the private sector.[20]

Unlike districted fleets of municipal haulers, private businesses can choose whichever company they'd like to haul their trash, escaping zoning regulations and thereby leaving haulers to operate with no efficient routing system throughout the boroughs. With dozens of competing companies servicing different businesses on the same streets, the 273 different private hauling companies drive a combined twenty-three million miles a year.[21] For comparison, if the DSNY drove the entirety of the city (six thousand miles) every day, it would total slightly more than two million miles a year. Even factoring in snow plowing and street sweeping, private hauling dwarfs the mileage of city-run operations. Driving millions of extra miles adds to carbon emissions, poor air quality in the city, and wear and tear on the trucks and their tires.

Picking up the waste of 8.5 million people is a huge undertaking, and there are some serious issues, but all in all it's a shockingly effective system despite it looking completely chaotic. With a population so notorious for hating change, it'll be an uphill battle to make adjustments for the sake of sanitation, visual tidiness, and fulfilling Eric Adams's personal vendetta against rats. But at least now the residents are battling against trash bins rather than stepping through tons of horse shit in the street.

Part 2

WHERE DOES OUR TRASH GO?

Chapter 4

POVERTY VS. WEALTH

T rash is handled differently around the world for myriad reasons, and it's important to address the disparities in garbage management before we get into the pros and cons of different disposal methods. There will be a fair amount of information in this book relating to the state of garbage around the world and how money, history, and politics affect it, so before delving into this section, I want to note that the terms *developed* and *developing* are being used less frequently when discussing a country's economic status or stage of growth. The concept implies there is one end goal and that all countries are or should be heading toward it, which isn't realistic, possible, or desirable. We don't need 195 Finlands! Dichotomies and binaries have historically not served us well; they inherently create opposition and hierarchy—us versus them.

Like all other political and cultural realms, there isn't a universally agreed-upon terminology for this concept of stages of growth in a country or divisions among them, and by the time you're reading this book, they may have changed again. It's all too complicated to sum up in a term that can describe half of the world. Countries are not unilaterally shifting on the same timelines or pace as one another, and within countries, economies and social freedoms can be wildly disparate from one region to another.

Global South and *Global North* are currently more accepted as decent, neutral terminology for describing very general patterns of global health, economy, and advancement (however that may look in that region) along geographical lines.[1] There are general patterns of differing poverty, health, human rights, economic disparities, and democracy between the Northern Hemisphere and the Southern. The Global North tends to be wealthier with more social and political freedoms, access to healthcare, income equality, infrastructure, and manufacturing. It includes but is not limited to Canada, the United States, Europe, South Korea, Israel, and Japan, as well as some countries that are actually south of the equator such as Australia, South Africa, and New Zealand.

The Global South, which encompasses most of Africa, Latin America, much of the Middle East, Asia, and the Pacific Islands, usually has newer democracies with greater levels of poverty, political upheaval, and a reliance on exporting raw materials, by way of mining or agriculture. Strongly rooted in a history of colonialism, economic divisions amongst countries around the world are largely a result of imposed limitations to grow and be independent players in the global system.

So, why did you need to slog through that semantics detour to get to the trash part of this? In an effort to succinctly discuss some of the topics related to global issues of poverty and garbage, I will do my best to use specific country names when possible—and when that's too clunky or will be a list of fifty, I'll use *Global North/South*. In the cases where I use *developed* and *developing*, it's in reference to specific realms of economy or social growth, not a holistic label of the country's state of being.

Now, on to the world's garbage.

Wealthy = Wasteful

The world's wealthiest countries produce the most trash—by a lot. The current statistics of waste production don't always show the complexity of why garbage production is high or low and how that waste is being dealt with. The curve of waste generation is a little bit wonky, factoring in the rate at which economic growth is happening and how long countries have had to establish systems of disposal.

Some islands, such as Tilos in Greece or Apo Island in the Philippines, produce almost no waste, usually due to a lack of access to plastic or non-biodegradable products, a holistic use of available products through reuse, and/or a need to keep consumption in check because of limited space to manage trash. Wealthy citizens in wealthy nations don't need to reuse goods for financial purposes; food, clothes, or excess *stuff* can be tossed with minimal thought. The value of waste lessens the more wealth is accumulated. Food waste isn't precious when there's always going to be enough resources to feed oneself.

The correlation between money and waste generation makes sense; it's simple in its basic premise and complicated in a million more intricate ways involving geopolitical factors. The more economic spending power a country has, the more it can produce or import. And the more economic spending power an individual has, the more they can consume. Because countries progress on such individual timelines from one another, there's a wide range of waste-generating norms taking place globally.

As many countries in the Global South are beginning to have economic booms that can aid the quality of life of their citizens, many of them are simultaneously experiencing drastic increases in pollution and waste problems. As citizens gain more access to wealth through increased opportunities in work and education, their ability to purchase newly imported or domestically manufactured goods expands. Rapidly industrializing nations, such as India, are experiencing a boom of economic growth, and therefore growth of waste with little government preparation or intervention. These nations are becoming some of the newest top waste generators.

Many countries with long-term financial wealth are starting to enter a new phase. After decades or centuries of increasing consumption and disposal infrastructure, they are now making efforts to use less and dispose of waste more efficiently. Spurred by climate change and a growing amount of waste to deal with (not to mention demands of citizens), countries have begun working to reduce waste through design—both of products themselves and cradle-to-cradle systems. Much of western Europe has passed legislation, modernized facilities, and implemented social education around multistream waste disposal. Western Europe is often viewed as the model of best practices when it comes to social and government programs—good healthcare, sanitation practices, worker rights, and transportation. It's wonderful to have models of successful social programs that are made possible by time and money; their economic expansion happened long enough ago (bolstered by colonialism), and their population is small enough to develop systems that accommodate their needs. Because of the spatial confines of small countries in places such as Switzerland and the Netherlands, they have designed more technologically advanced incineration facilities than the United States or China, who have more than enough space for sprawling landfills.

Over the last several decades, China's economic boom threw it into an environmental crisis, surpassing the United States as the top emitter of greenhouse gases. However, it should be noted that much of that comes from manufacturing, leaving the US at a per capita emissions rate twice as high as China.[2] China's biggest landfill, Jiangcungou, filled up four times faster than planned for—a glimpse into the overwhelming amount of garbage being produced from rapid economic development. A different landfill, located in Shenzhen, was the site of a landslide that killed seventy-three people in 2015.

CLIMATE REPARATIONS

The United States, despite a history of wealth, has long ignored the trash problem and continues to practice behaviors that we, at this point, recognize are harmful. But legislation has, for the most part, not stepped in to curb our habits. The ways we deal with waste reflect income inequality and racism, not just in how and what we can purchase but in how our waste is disposed of and who is affected. As awareness grows about the global state of garbage, manufacturing, and its ill effects, it's becoming clear how that has affected countries who are most at risk from climate change. Calls for "climate reparations" are being made. A study of the impact wealthy countries have on global emissions found that the United States "has caused more than $1.9 trillion in climate damage to other countries from 1990 to 2014."[3] As extreme weather conditions are happening more and more frequently with greater intensity, countries that are experiencing the most damage are often the ones who emit very few greenhouse gases and produce very little garbage. After massive flooding in Pakistan in 2022, greater attention was brought to the growing need for wealthy countries and private companies to compensate for their part in causing climate change. Our exported trash pollutes the water and soil in the countries saddled with our excess, and poses health risks to the individuals who make a living disposing of it. The responsibility is ours, even if we can't see the damage we're causing.

Racism, Pollution, and Policy

The US is usually lumped in with Europe and Canada as part of the Western world in areas of economic, social, and political status, and there is immense privilege for many who live here. But unlike much of Europe and Canada (which are by no means perfect), the United States does not handle its social welfare, economic or racial equity, environmental protections, or waste management in ways that are sustainable or equitable.

There is an eighty-five-mile stretch of the Mississippi River between Baton Rouge and New Orleans that is known as Cancer Alley. Residents there are used to their friends and family members dying early, or their children developing chronic asthma. This corridor along the river is home to over two hundred heavily polluting chemical factories, industrial plants, and oil refineries with a population that is majority Black and low-income.[4] These residents face a form of environmental racism that is apparent throughout the country (and the world) in staggering numbers. Rubbertown in Louisville, Kentucky; Flint, Michigan; Uniontown, Alabama; Harrisburg in Houston, Texas; the list goes on.

The term *fenceline community*, also known as *frontline community*, generally refers to a population within three miles of high-risk chemical facilities, natural gas/oil refineries, and other toxin-producing manufacturing. There are more than twelve thousand of these sites in the US, the demographics of which makes it very apparent where they're clumped, or rather, who is clumped around them. Fenceline communities bear the brunt of industrial pollution, facing disproportional amounts of exposure to air and water pollution as well as an inordinate concentration of cancer-causing chemicals. With these factories also come other forms of pollution that, while not toxic, are detrimental to the health and well-being of nearby residents, including noise and light pollution.

As of 2023, the United States population is 13.6 percent Black, but 78 percent of Black individuals live within thirty miles of a coal-fired power plant and breathe 37 percent more nitrogen dioxide than white people—a relic of segregation and redlining.[5] In New York City, the South Bronx, Southeast Queens, and North Brooklyn all bear the brunt of trash collection disturbances and related health issues. A report compiled by the organization Transform Don't Trash NYC found that the South Bronx is home to fourteen transfer stations (23 percent of the city's stations), with 304 commercial trucks passing through every hour, amounting to more than 50 percent of the commercial waste trucks in the city. Residents there suffer from two to seven times more asthma-inducing pollutants than the national average. Similarly, North Brooklyn is home to 38 percent of transfer stations with 203 trucks passing through per hour and five times the rate of asthma-inducing pollutants.[6]

Each of these communities in New York is made up of primarily Black and brown people experiencing higher rates of poverty than their surrounding neighborhoods, such as gentrified Williamsburg. These stations impact quality of life and mental health as much as physical health, with constant disturbances from loud trucks at all hours, eyesores, and bad odors. There is a history of fighting against environmental racism by those in affected communities, such as the 1978 class action lawsuit *Bean v. Southwestern Waste Management Corp.*, in which residents of the predominately Black neighborhood of Northwood Manor sued for environmental discrimination against the City of Houston's proposal to build the Whispering Pines landfill (a very pleasant name for a very unpleasant facility).

In that case, Dr. Robert Bullard was hired to conduct a study of the waste disposal site, which was located dangerously close to public spaces and represented a blatant disregard for the health and well-being of the neighborhood's residents. He stated, "It was the height of disrespect compounded by the fact that the landfill was 1,300 feet from a high school in a Black school district and with at least a half-dozen elementary schools in a two-mile radius."[7] The residents lost their case and the landfill was built soon after, followed by several other industrial facilities. Despite the loss, Bullard continued his work in the study of environmental racism, and because of his important work in the emerging field, Dr. Robert Bullard is regarded as "the father of the environmental justice movement."

Following the *Bean* case, an extensive study of national environmental racism was spearheaded by Rev. Benjamin Chavis Jr., a North Carolinian who led the United Church of Christ Commission for Racial Justice. His study made clear that there was a strong correlation between race and proximity to toxic waste sites and landfills, with three-fifths of Black and Hispanic Americans living near these types of facilities. While many com-

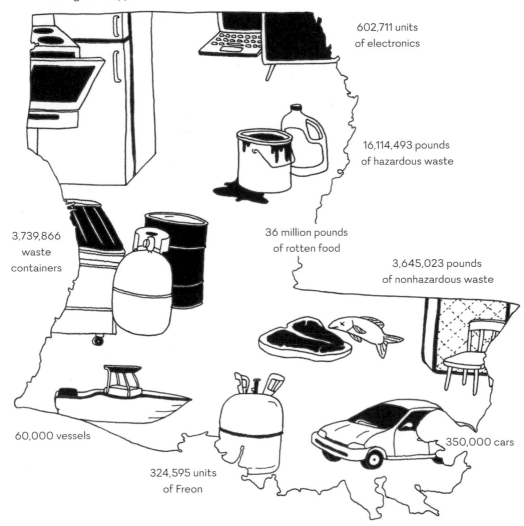

891,996 units of "white goods" (appliances)

602,711 units of electronics

16,114,493 pounds of hazardous waste

3,739,866 waste containers

36 million pounds of rotten food

3,645,023 pounds of nonhazardous waste

60,000 vessels

350,000 cars

324,595 units of Freon

munities affected by environmental racism in the United States have high rates of poverty, one journal found that "African-American families with incomes of $50,000 to $60,000 were more likely to live in environmentally polluted neighborhoods than white households with incomes below $10,000," indicating that above all else, race was the most significant factor in likelihood to be close to toxic facilities.[8]

Over the last fifty years, residents and activists have banded together to fight against these facilities being built and operated in their neighborhoods. Some have been successful, with Superfund sites being established to clean up the area. Some were not, and landfills and refineries either remained operational or were built despite community opposition, continuing to cause cancer, asthma, and other illnesses for nearby residents.

In Philadelphia, Terrill Haigler has garnered media attention online under the moniker "Ya Fav Trashman" for his work advocating for increased city investment to eliminate illegal dumping in Black neighborhoods. "Trash is really an attack on the disinvested," and it's a reflection of a city's disregard of its own population that can perpetuate racist neglect and ideas of class.[9] Providing less governmental care to more disadvantaged residents of a city is par for the course for America. Studies have shown that reducing the amount of garbage and debris from municipally neglected areas like streets and vacant lots can improve crime rates, encourage the use of outdoor social spaces, and boost the mental health of those in the neighborhood.[10] It makes sense; seeing your space be cared for and respected makes you feel respected—and the opposite is also true.

This dynamic of economically driven trash disparity is not isolated to the United States; it's rampant in other countries where waste management is informal and unregulated. When countries urbanize rapidly—with individual spending power and manufacturing output both increasing—trash production grows, too. This pattern can be seen across a wide array of environmental issues, such as worsening air pollution from increased use of fossil fuels, increased water consumption for industry, and ill-equipped waste management systems.

Those Behind the Scenes

The global differences in waste disposal systems are complicated and profound. In much of the world, waste disposal is an unregulated system run by individuals and groups with little oversight. Around twenty million people worldwide make their living as waste pickers, sorting and recycling the world's garbage.[11] In many parts of the Global South, they are the primary infrastructure of recycling—creating microeconomies that sometimes become contracted with local municipalities. While waste pickers are an invaluable part of economic and environmental systems, they are largely looked down upon and without legal worker protections.

Waste pickers serve an important purpose in many countries that don't have an established waste management system in place. Landfills are the most common method of disposal in low-income countries, as they're much cheaper to establish and maintain than incinerators; 90 percent of the garbage in low-income areas is openly burned or disposed of in unregulated landfills.[12] Municipal waste systems require immense funding to establish and operate, and as countries rapidly urbanize, the rate of garbage generation is increasing with no system in place to meet higher demands. Thanks to the unpaid labor of waste pickers, cities have emptier landfills (still very full, but emptier), cleaner public spaces, less pollutants reaching waterways, more money in the economy from recyclables, and less need for virgin materials. And, for the most part, cities get these benefits without paying a dime.

Because of the world's lack of ability or desire to adequately care and provide for those in poverty, many must turn to waste picking to survive, or are born into multigenerational families of waste pickers. Waste provides income, with massive dumps creating work for people living in poverty who may not otherwise have any job prospects.

The quality of life among waste pickers can vary. Some are on the edge of survival, earning next to nothing and living on site at the dumps in makeshift shanties with other waste pickers; sometimes hundreds of people live in these informal communities with their children, who begin to pick waste at a very young age. Others make fairly good wages, finding the more valuable materials to sell to local middlemen, to businesses, or straight to the government. Waste pickers provide an essential service to cities that have no formal recycling or municipal trash systems, with a significant portion of waste collection in the Global South being performed by waste pickers.[13]

There are several types of collection within the scope of waste picking: autonomous, cooperative, or contracted. Within each of these, there are many types of jobs one can do, depending on what types of waste are in the area, how accessible they are, and whether someone works in a residential or commercial setting. Independent waste pickers operate autonomously, individually collecting recyclables or organic matter from landfills or public areas to sell to middlemen or junk shops, some of which specialize in specific materials, such as cardboard, metals, or glass. These individuals have no official affiliation but are often part of communities with other pickers, working and/or living in the same area. Organizations and cooperatives often do political work to increase worker rights for the profession, and in certain places, Brazil and Colombia in particular, have found success in gaining recognition and rights from governments.

John Chweya, a lifelong waste picker at the Kachok dump in Kenya, describes waste pickers as "the backbone of collection and recycling systems in the world."[14] He has been instrumental in collectivizing Kenyan waste pickers to have more visibility and negotiating power. After years of mistreatment as a waste picker, Chweya created a cooperative that is able to negotiate and earn better wages. He is one of many people who are organizing campaigns initiated by workers, which is often the only way to do it—efforts are rarely taken by the government to protect those doing the dirty work at no cost to them.

Sometimes collectives or individuals can become formally recognized by the government to work for junkyards, municipal waste sites, or cooperatives—receiving wages and legal employment status. In Buenos Aires, Argentina, waste picking became an organized government system when a law banning waste picking was reversed in 2002. Paired with an increased effort to clean up city streets and reduce landfill usage, a niche opened up for waste pickers to fill. Pickers, known in Argentina as *cartoneros*, formed cooperatives that negotiated with the city government for a monthly salary to pick up recycling. While the livelihood of waste pickers improved significantly with the stability of salaries, they are still paid far less than the private recycling sector.

Many waste pickers take pride in their skilled labor, but receive very little recognition and are exposed to a myriad of severe health and safety

SAN FRANCISCO'S → $550,000, 4-YEAR ←

QUEST TO MAKE A PERFECT, OVERPRICED CITY GARBAGE CAN

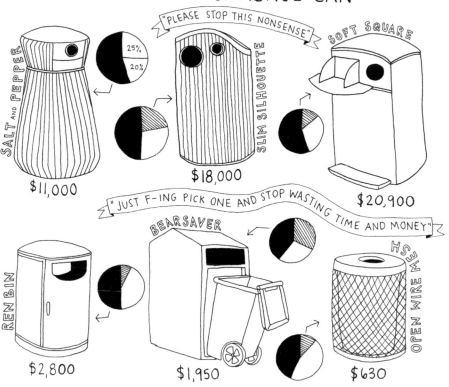

"PLEASE STOP THIS NONSENSE"

SALT AND PEPPER
25%
20%
$11,000

SLIM SILHOUETTE
$18,000

SOFT SQUARE
$20,900

"JUST F-ING PICK ONE AND STOP WASTING TIME AND MONEY"

REN BIN
$2,800

BEARSAVER
$1,950

OPEN WIRE MESH
$630

■ NOT IMPRESSED AT ALL   ▨ IT'S OKAY   ▧ LOVE IT

hazards. Interacting with open landfill materials without proper protection can be extremely dangerous. Sharp objects or falling debris can lead to infected cuts and injuries that can prevent continued work or even lead to permanent disabilities. Diseases and infections like tetanus, hepatitis, and parasites are common among waste pickers, and longer-term health effects like cancer and endocrine issues affect workers and residents at the landfills. Open-air dumps and burn piles almost always have negative impacts on their surrounding communities (human and otherwise), and even the most sophisticated methods of disposal have their issues with establishment, maintenance, and shutdowns.

Whether you're a waste picker at a landfill or a soccer mom in the suburbs, you have a part to play in the world's trash ecosystem. Someone always has to deal with the waste we generate, and depending on who and where you are, your experience of garbage can wildly differ. Wealth and race play a big part in your relationship to trash and how it affects your way of life.

THE CITY OF TORONTO, CA, SPENT $31 MILLION TO CREATE 500,000 NEW RACCOON-PROOF TRASH BINS...

WHICH SOME RACCOONS CAN OPEN IN 30 SECONDS

SULFUR-CRESTED COCKATOOS IN SYDNEY, AUSTRALIA, HAVE TAUGHT ONE ANOTHER TO OPEN TRASH CANS

WITHIN ONE YEAR, THE BEHAVIOR SPREAD FROM 3 TO 44 SUBURBS

The trash can design problem.

Chapter 5

RECYCLING

n 1987, a barge started a lonely six-thousand-mile journey of rejection that would change the face of recycling. Loaded with 3,186 tons of garbage, the Mobro 4000 barge was set to haul Long Island's trash to Morehead City, North Carolina, where it would be offloaded into a landfill to create methane-generated electricity. Lowell Harrelson, a farmer from Mobile, Alabama, hatched the plan as a method of making money from hauling waste and selling electricity to the power companies—an idea ahead of its time that's now common practice, with 68 percent of US landfills generating electricity.[1] At the time, New York City was eager to ship away its ever-growing trash problem, and Lowell was eager to try this idea in hopes of creating a profitable business.

Harrelson needed a dock to launch from; for that, he turned to Tommy Gesuale, a private dock owner. For capital to back the endeavor, he asked Salvatore Avellino Jr., a mafia boss affiliated with the Lucchese family that controlled the waste hauling industry on Long Island. The mob at that time paid to dump their garbage in New York landfills, so Lowell presented an appealing, cheaper option of hauling it for them and avoiding the landfill fee altogether. It was the start of an enterprise that Lowell hoped would provide a new way of dealing with garbage and creating energy.

Barges carrying loads of garbage weren't a common sight up and down the East Coast, so once the barge arrived on the coast of North Carolina, it became a media sensation, garnering crowds of onlookers and the eye of an environmental department employee who, after seeing a bedpan in the tons of waste, suspected that Mobro 4000 was hauling illegal medical waste. After investigating, it was determined that there was no hazardous waste being carried, but the media sensation around the barge had already grown too big. After being turned away from unloading in North Carolina, the barge was subsequently denied access to landfills in Louisiana, Mexico, Belize, and Key West. After five months of rejection after rejection, the Mobro 4000, by this point deemed the "gar-barge," docked back at its starting point on Long Island.

After facing myriad obstacles in the way of converting waste into electricity, the Mobro 4000's load was ordered to be incinerated, not landfilled. Once workers began unloading the garbage to be burned, it was discovered that a significant portion of the waste was recyclable materials, inciting anger amongst activists that the materials should have been diverted to recycling rather than ending up on a barge of garbage. The barge became a symbol of the growing amount of trash being produced in the United States—a visible mound of waste floating on the ocean that was the center of media attention for months; the antithesis to "out of sight, out of mind" that had, in part, created such a mess.

A contributing factor to the outrage and rejection of the barge was a misconception that landfills were reaching their capacity and space was becoming increasingly scarce. Landfills in the 1970s and '80s were closing by the thousands, leading people to believe it was due to being at capacity and no longer able to accept waste. However, the reason behind the closures was not lack of space, but rather new, stricter laws prohibiting certain practices in older landfills, such as not having liners to protect groundwater from leaching waste chemicals, known as leachates. While local open-air dumps were closing, bigger EPA-approved landfills were being built, creating acres of space for millions of tons of garbage.

While the Mobro 4000 appeared to have an unfathomable amount of waste (more than anyone normally sees), the equivalent of seven Mobro 4000s is shipped out of New York City every day. Because of the environmental awareness spurred by seeing that amount of trash, in conjunction with the panic of landfills closing by the thousands, people felt more compelled to recycle their garbage in an effort to reduce the amount of material entering the landfills—under the misconception that they were nearly full. Recycling had been picking up in popularity in the decades preceding the gar-barge, but municipal recycling took off in earnest following the media spectacle it created.[2]

The Root of Modern Recycling

A hundred years before the barge incited anger and calls for action to re-cycle, the foundation was laid to create a strong push away from recy-cling and toward mass waste generation. Both are ultimately a win for manufacturing—one just a little more obviously so. The general ethos of recycling before the turn of the century was far more communal and wholly practical; a smaller population relied on bartering economies to circulate valuable materials between producers and consumers. With a more limited availability of materials sourced from the natural world, the system of recycling in many places was an effective closed loop with very little waste, much like "sustainable biological ecosystems, which are in general closed, or cyclical. Waste to one part of the system acts as re-sources to another."[3]

In order to sustain households, businesses, and manufacturers, there was incentive for consumers and producers to rely on one another to cir-culate materials and goods. A big player in the early development of re-cycling systems was rags: most likely not something that's on your mind most days. We clean with them, toss them in the laundry, the trash, or a gross pile in a corner, and unless rags are part of your job (which they are for many people), they're an unremarkable part of our lives. Rags in the 18–1900s, however, were a central part of the emergence of a recycling economy—valuable to people and businesses alike. Because consumerism was yet to exist as we know it, and repair was the central way people maintained and reinvented goods, textiles were used as many times as possible before becoming unwearable; and once at that stage, they were turned into rags for extended use.

At the time, paper wasn't made from wood pulp but from cloth, mak-ing rags a valuable resource to paper mills who were willing to pay citi-zens and businesses for their scraps. Peddlers were often the middlemen, collecting rags and other recyclable materials in exchange for money or goods to sell to businesses or manufacturers. The transition from a recy-cling economy as a means to mutually acquire goods (circularity between consumers, businesses, and manufacturers) to a linear consumeristic one (manufacturers to consumers to disposal) began when the incentive to get rid of reusable material was greater than to thrift and repair. After losing access to Britain's paper supply upon entering the American Revolutionary War, the United States met increasing paper demands to fulfill domestic publishing in the 1770s by means of a government cry for citizens to recycle their rags in the name of patriotism.[4] People followed orders, and rags became less for the household and more for papermaking.

Bottle Recycling

At the same time that rags were transitioning into a linear commodity, glass was still very much circular. Bottle making was an expensive, time-consuming process that incentivized businesses to reuse glass bottles as many times as possible. Since glass doesn't degrade in quality with continued use, it can be reused an infinite number of times.

Restaurants and bars kept bottles, refilling them in-house. When customers bought beverages in bottles, they could return them to the store from which they were purchased; and to make up for the up-front cost and short supply, sellers offered discounts and rebates to those who returned their glass bottles after use or brought their own to fill.[5] Some establishments branded their bottles to retain their ownership so customers would return the bottles rather than selling them to secondhand bottle buyers who were inching out small exchange markets in favor of large-scale collection and redistribution enterprises. In addition to secondhand bottle

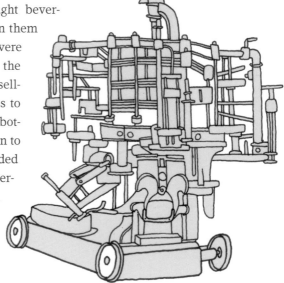

First mechanized bottle maker, 1903.

Invented by Michael Owens and introduced to the commercial market in 1920, it had 10,000 parts and weighed 30 tons.

It could make 240 bottles per minute.

markets, bottle exchanges cropped up, serving as a hub to return bottles to their original businesses.

Bottles stayed in circulation, rarely being discarded until the advent of the mechanized bottle maker, invented by Michael J. Owens in 1903. The bottle making machine wildly outpaced hand making them, with the ability to produce 240 an hour. As we've seen in the case of most mass pro-

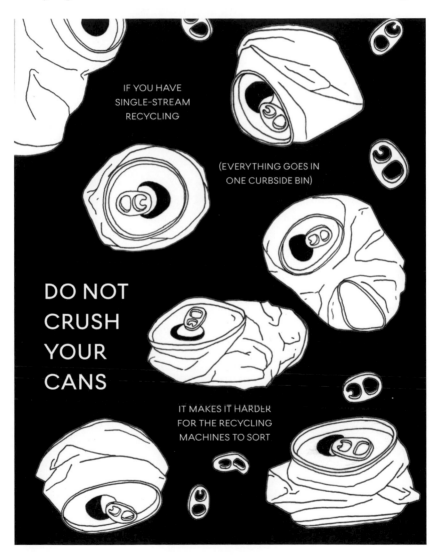

IF YOU HAVE
SINGLE-STREAM
RECYCLING

(EVERYTHING GOES IN
ONE CURBSIDE BIN)

DO NOT
CRUSH
YOUR
CANS

IT MAKES IT HARDER
FOR THE RECYCLING
MACHINES TO SORT

duction, the more readily available and cheap something is to make and buy, the more often it ends up in the trash.

By the 1960s, cans replaced a significant portion of glass, particularly in the beer industry. Sodas continued to be bottled in refillable glass containers that came with deposits: a great incentive to encourage people to reuse. Even if it's only a few cents per bottle, we love to get our own money back. The deposit system faded out of favor after World War II, and by 1970 more than half of beverages were sold in cans and bottles that had the same infinite lifespan but would end up in the garbage after a single use. Retailers no longer wanted to deal with collecting an increasing amount of bottle returns, and as more people took drinks home, they had no easily accessible disposal method, as curbside recycling didn't begin in the US until 1981.

Between 1970 and 1986, ten states and all of Canada signed on to enact "bottle bills" that required all beverage containers (both glass and cans) to have refundable deposits. Ranging from five to fifteen cents each, bottle refunds have led to significantly higher recycling rates in states that offer them. Michigan leads the pack with a redemption rate of 89 percent at ten cents apiece.[6] The ten states with implemented bottle bills consistently produce a higher rate of recyclable materials, and because bottle bills create systems that inherently separate materials, they provide a necessary, streamlined source of materials to maintain postconsumer PET (a type of plastic used to make beverage bottles, peanut butter jars, and other common food packaging) and glass production. The more sorted recyclables are when entering the system, the less time and labor is required at the facility.

Cans and bottles are turned in at recycling facilities, bottle redemption stations, some gas stations, and local grocery stores. There's no limit on the window of return for deposit cans and bottles, allowing them to be dropped off at any time after purchasing. Sometimes people hold on to their bottles a little too long. For example, in Germany in 2022, the Ukraine war put a pause on imported glass bottles, creating a shortage for brewers—of which there are over a thousand in Germany. And Germans love beer. Even with all the bottles stashed away at people's homes, Germany is still the world's top recycler, coming in at a recycling rate of 56 percent of all waste.

If you live in a bigger city in one of the states with bottle bills, you probably know that some people, known as "canners," make their livelihood collecting bottles and cans from residential or commercial bins or litter on the street. If recycling cans or bottles doesn't feel like a worthwhile use of your time or financially incentivizing enough to follow through on, consider collecting and placing your recyclables where they can be picked up by someone who relies on the redemption money to survive.

A Brooklyn-based organization, Sure We Can, works to aid canners' livelihood, improve their living conditions, and foster community with other canners. The organization has physical space for canners to store and exchange their collected goods, as well as social and educational programming, including a community garden and classes. SWC works to bring awareness to the invaluable role that canners serve in the community, and currently works with 740 canners who collected eleven million cans and bottles in 2021.[7]

The beverage industry is not a fan of the bottle bills, as they work like a tax for it. Governments keep the money from unreturned bottles and cans, while beverage companies shoulder the cost—rather than taxpayer-funded municipal recycling or disposal systems. Lobbyists and government officials have fought tooth and nail to keep bottle bills from passing into legislation, keeping recycling rates in those forty states much lower than the ten who have them. Beverage companies in particular push the idea that recycling is the key to a healthier environment, and an action that consumers can do for the common good. Plastic producers realized early on that "selling recycling [sells] plastic" and in order for them to continue increasing production of single-use plastic packaging, consumers had to believe in the wonders of recycling—a guilt-free pass to purchase garbage.[8] The world of recycling has been made opaque in a campaign to place the responsibility on consumers to make a difference—a tactic that comes directly from advertising by corporations who produce products that are (in theory) recyclable.

Sometimes campaigns calling for individual behavior changes have incredible and unexpected success, such as the "Don't mess with Texas" campaign. The phrase was introduced in 1985 with the intention of targeting men ages eighteen to twenty-four, who were identified as the primary source of littering in the state. Delivered in a more casual tone than your normal Department of Transportation signage, the message was first promoted by Stevie Ray Vaughan in 1986, and since then has been supported by many other Texas legends, such as Willie Nelson and George Strait. The phrase has since become synonymous with Texas in popular culture.

It All Comes Down to Marketing

In 1953, the "Keep America Beautiful" campaign began as a public call to stop the littering problem, and start recycling as a way to save the planet. Under the guise of environmental stewardship, KAB was funded (behind the scenes) primarily by the beverage, tobacco, and packaging industries, with Coca-Cola and Dixie being the most well-known brands in the bunch. Their funding of the campaign was not an altruistic attempt to offset their production of goods, but rather to misdirect feelings of responsibility away from themselves and onto consumers. In one famous ad, the message is "People start pollution. People can stop it."

TRASHY TIDBIT

Roosevelt Island and Walt Disney World

What do Disney World, Roosevelt Island, Stockholm, Tel Aviv, and Mecca have in common?

Their trash gets sucked away underground at sixty miles per hour.

Disney World doesn't naturally conjure up a picture of environmental innovation, given that the corporation isn't historically known for its ethics or environmental consciousness (in part thanks to Disneyland's annual ninety thousand pounds of fireworks).[9] However, in its efforts to keep the parks unnaturally clean, give the illusion that there are no maintenance workers or out-of-place characters roaming around outside their fictional world, and maintain the illusion of utopia, Disney installed the first pneumatic trash system in the United States. Automated vacuum collection systems (AVACs) line the tunnel systems of the Magic Kingdom, sending garbage every fifteen minutes to a compactor behind

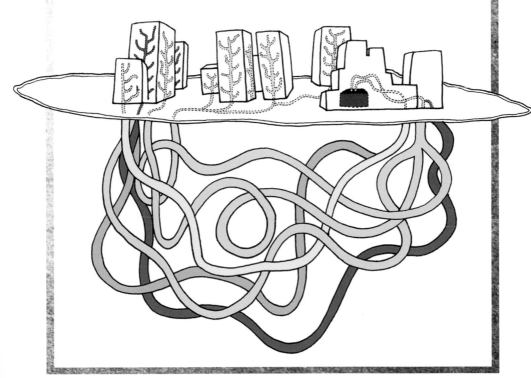

Splash Mountain that works to crush the garbage of more than twenty million visitors per year.*[10]

Soon after, in 1974, Roosevelt Island in New York City opened its pneumatic trash system, which handles all garbage generated by its twelve thousand residents.[11] The underground tubes vacuum away garbage from residential chutes at seventy miles per hour to a central location where it's sorted, compacted, and shipped off the island. This system eliminates the need for garbage trucks and keeps rat-attracting trash bags off the street, creating a much more time- and energy-efficient method of disposal.

With only six dials and no computers involved, the analog system hasn't needed much updating to remain just as functional fifty years later. The issue that has evolved is not with how well the machines work, but what has been tossed down the chutes. Just like recycling has become a free-for-all of what's seen as bin-worthy, so has the garbage on Roosevelt Island. Items such as Christmas trees and furniture warrant Swedish crews (the experts of the AVAC world) to come in and free the garbage from its stuck state.[12]

*This number reflects average attendance before COVID-19 shutdowns.

In direct opposition to their public messaging, when bills were proposed to curb the production of plastic packaging, the companies behind the face of KAB staunchly opposed the bills in order to retain the rights to produce products in single-use plastic bottles and other nonrecyclable containers.[13]

As the public became more at ease consuming massive amounts of plastic without landfills weighing down their conscience, the more varied plastics became. While plastics remain largely made of virgin materials, using either oil or natural gas, they are often not a single type of plastic but an amalgamation of many plastics that are close to impossible to turn into high-grade usable material.

The current state of recycling's effectiveness is hotly debated, but overall, it's a pretty grim scene. In theory, recycling is a good way to close the loop on material sourcing, but in practice, the economic structure of functional, large-scale recycling isn't in place to make it financially incentiviz-ing to companies or governments; raw materials are cheaper and easier to use than recycled material, and the profitability of plastics is enormous—raking in $400 billion per year for the oil industry.[14] In addition to the inadequate infrastructure, the public is poorly educated in proper sorting, cleaning, and purchasing, which isn't helped by convoluted and mislead-ing packaging from manufacturers.

Recycling in Practice

I live near the US's largest creative reuse center, and I often take in boxes of things that are most certainly not up to snuff for a thrift store (a half-used ink pad, tiny vials, a handful of mismatched buttons, cords from unknown source, scraps of fabric), but they have life left in them—either in their original form or as material for transformation. I know the reuse center will keep a significant portion of it to sell to people who, like me, go to look for something specific and leave with a box of miscellaneous objects and materials that have potential to be creative projects or useful items, like a half-used roll of packing tape. I generally have a junk bag in my car that I forgetfully drive around with for months before actu-ally dropping it off, even if I've collected new junk from them in the meantime. It feels important to bring items there that I know would be deemed garbage by any other secondhand store and aren't quite garbage enough to throw away.

Had I brought my random buttons or old envelopes to a Goodwill or other local thrift store, they would certainly be thrown away, never mak-ing it to the sales floor. I would be succumbing to the shiny, hopeful con-cept of "wish-cycling": the belief that something is recyclable or usable that will actually end up in the garbage and impede the recycling process. Throwing nonrecyclable goods into private or municipal bins is extremely common—a product of being poorly educated on the proper ways of dis-posal, plus a lack of good options for things that don't fall in the purview of collectible recycling. Wish-cycling can disrupt recycling processes or render whole batches nonrecyclable, so it's important to follow guidelines of your local municipality when throwing things into the bin.

AN ISLAND DOING IT RIGHT

On Shikoku Island, Japan, a town of 1,500 residents developed a system of recycling and reusing that has brought their town 80 percent of the way toward zero waste. What started as a nine-category sortation system in Kamikatsu has evolved into a forty-five-category system housed in the Zero Waste Center (ZWC): a building made largely from reused materials, including seven hundred donated windows that fill the space with light.

There is no municipal garbage collection, but a free service is provided to pick up household trash from those who can't transport it to the facility themselves. The ZWC serves as a drop-off center where residents bring their waste and carefully sort all their materials into their respective areas. If you tried to list out forty-five categories that waste can be separated into, would it include heating pads, toothbrushes, mirrors, lighters, ballpoint pens, or mercury thermometers? Those, as well as paper cups, broken fluorescent lights (separate from intact ones), razors, cosmetics, plastic caps, and chopsticks are all some of the many materials collected and sorted. Much of the recyclables generate income for the town, offsetting what needs to be sent to landfills or incinerators outside town.

On the same property is the Kuru Kuru shop (a thrift store of sorts), where all the goods are free, so long as they're weighed on the way out. The weight is logged by the shopper in a paper ledger to keep track of the volume that's come in and out of the space—a record of the success of reuse. The most recent addition to the center's grounds is a four-room hotel, the Hotel Why, that brings zero waste to the forefront of their guests' experience—cutting off only as much soap as is needed using a cheese slicer, sorting any trash created into six (out of the forty-five) categories, and gaining access to the locals-only recycling facility. The space is beautifully designed; a testament to the potential of low-impact design, it was built with reused and salvaged materials, including secondhand furniture and blankets made from scrap textiles.

A primary reason we're prone to wish-cycling is ignorance of how it all works. We've been conditioned to look for the triple arrow symbol we all associate with the purity of recycling, which has ultimately exacerbated the issue, making it nearly impossible to enforce an effective system. Popularized in 1989, the universal recycling logo began appearing on plastics of all varieties. The symbol elicits our trained *this can be recycled* response, and into the recycling bin it goes. I doubt most of us pay attention to the numbers (1 to 7) within the symbol on the bottom of our plastic containers, and probably even fewer of us know what those numbers actually mean. The truth is that most municipal collection systems only accept numbers 1 and 2. The ubiquity of this label, like the Keep America Beautiful campaign, was pushed for by oil and plastics lobbyists in the late 1980s in an effort to further escalate the public belief that recycling was both easy and functional; if recycling is encouraged more and more, it must be because it's working. So how do we move away from ineffectual recycling caused by flawed systems of manufacturing?

Just Don't Use It

Our flawed thinking around waste is perhaps best exemplified in the straw issue. I'd hazard a guess that almost no one is routinely using straws at home for things other than a milkshake. For most able-bodied people, a straw is far from a necessity, and if the world stopped producing and providing straws for those who don't require them, we'd all be more than fine. The search for a paper or biodegradable alternative is much more of an undertaking than a simple shift in mindset: you probably don't need a straw. Ever. But people have come to expect certain products as part of certain experiences, so manufacturers have a continued demand, even if it's a passive one, to produce these items.

We have a culture of eating on the road, getting to-go coffee, and having mixed drinks with those little black straws that are way too small to drink out of and don't seem to serve much of a purpose. In order to shift our problem with waste, we have to shift our behavior and our expectations. I keep a spoon and fork in my car for when I might unexpectedly need them. Sure, they're disgusting and covered in dog hair, but they're easy enough to clean off when I need them. I suggest keeping your own somewhere in the car that's a little less open to the elements, but the principle stands.

There is a reason recycling comes last in "reduce, reuse, recycle," yet we've turned it around so it feels much closer to "recycle, reuse, reduce." Recycling only works if it holds value in the economy, which, since China imposed a ban on importing US recyclables, has plummeted domestically. The process is usually more expensive to operate than landfilling, often costing cities far more than they can get back when selling the recycled materials. The economics aren't sustainable, particularly since paper and plastic degrade with every round of recycling, with plastic often being made into carpet or other items that eventually end up in the landfill. Recycling is not the solution to the garbage crisis.

Inside a recycling plant

I hate to end on a note that the miracle cure isn't a cure at all. Some recycling is effective, but labeling something as recyclable doesn't make it inherently good or necessarily better than throwaway, single-use items. A heavy recyclable item is more energy-intensive to dispose of than a single-use, throwaway item that's lightweight and made from partially recycled materials. We need to be thinking of how to manufacture and recycle *smarter* rather than focus all our do-gooding toward recycling as much volume as possible. Sure, generating millions of plastic bottles that are technically recyclable produces the most recyclable material, but that doesn't mean it's the best option.

We need to move away from obsessing over recycling and move toward a system that implements, encourages, and enforces companies to design products that prioritize closed-loop thinking. How can materials remain functional use after use without losing quality? What can we adapt our lifestyles to need less of, and how can the economy support that through extended producer responsibility or consumer incentives?

Chapter 6

LANDFILLS
AND INCINERATORS

There are two main ways, outside of recycling, that trash is handled: landfills and incinerators.

Incinerators are sometimes referred to as "thermal treatments," which sounds like something recommended by Goop, but they're more commonly called "waste-to-energy facilities," or WTEs. Burning is an age-old, fast, and cheap solution if done without oversight, and an expensive option if built and operated with strict regulations.

The more common option is landfilling, or open dumping, which accounted for 70 percent of global waste disposal in 2018, followed by 19 percent recycled or composted and 7 percent incinerated.[1] Landfills are variable in their level of oversight, ranging from well-managed daily soil coverings and wastewater treatments to open dumps where toxins leach into the soil, release greenhouse gases, and allow for disease to easily spread—frequently contracted by waste pickers. They are also some of our largest man-made creations, often spanning huge swaths of land.

A landfill is the most intuitive and oldest form of waste disposal, dating back to the piles of pottery shards and shell mounds of ancient populations. Tossing things in a big pile is a natural impulse to deal with the things we don't actually want to deal with (dishes, laundry, garbage), and we continue to do as much. It's a place to dump things we don't want (or know how) to deal with. Any surface in my house is chronically a mini landfill.

HOW TO START A LANDFILL OR RECYCLING PLANT FIRE

(THIS IS NOT MEANT TO ENCOURAGE ARSON)

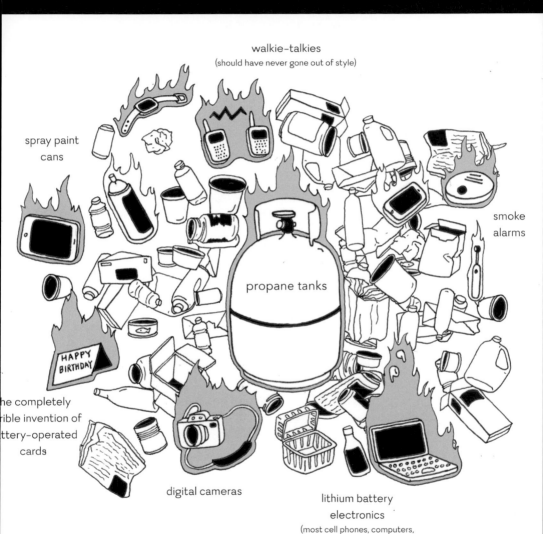

walkie-talkies
(should have never gone out of style)

spray paint cans

smoke alarms

propane tanks

he completely rible invention of ttery-operated cards

HAPPY BIRTHDAY

digital cameras

lithium battery electronics
(most cell phones, computers, AirPods, smartwatches)

So, Which Is Better?

The financial, human health, and environmental costs of both landfills and incineration are highly variable, making them hard to compare against one another. Across almost all realms of trash disposal, there are widely differing opinions about the merits and drawbacks of each method, and the debate surrounding landfills versus incineration is a major point of contention among environmentalists, waste industries, and governments.

There are misconceptions about both methods of disposal, with the most extreme of our notions being that landfills are always putrid, rotting pits, and that incineration is guaranteed to spew toxic ash. In many places that lack funds for high-tech, expensive facilities, the health repercussions of any forms of trash disposal are unsightly and potentially deadly, particularly when people make their living interacting with waste with very little protection, or when there are no protective measures for how burned materials are contained or filtered.

I spoke to Sam Hawes, landfill manager at the New Hanover County Landfill in Wilmington, North Carolina, to understand more about the relationship between incinerators and landfills, how they can be integrated, and public perception surrounding the methods of disposal. The New Hanover County Landfill is a well-established, multifaceted facility with its own compost operation, environmental programs, reverse-osmosis water treatment plant, and construction debris recycling program; and it's soon to begin an operation selling gas produced by the landfill to the regional natural gas company.

For twenty-five years, the company had a five-hundred-ton per day waste-to-energy (WTE) incinerator as part of its operations before it fell into disrepair. It was ultimately shut down by elected officials because the landfill was cheaper to operate per ton than incineration. It might seem counterintuitive to have both a landfill and WTE facility in one place, but the relationship was a symbiotic one between the two modes of disposal that, when combined, made for a highly efficient outcome.

Hawes explained that needing to regularly test for hazardous metals in the ash was one of the biggest issues in older plants and is responsible for closing many WTE facilities twenty-five years ago. The recent loss of the Hanover WTE facility has caused a big financial and supply strain;

federal EPA regulations require all landfills to cover their trash daily with six inches of material (generally soil) to reduce odor, minimize materials being displaced from wind, keep animals away, and physically crush rats already on-site. But because the incinerator was part of the Hanover operations, the incinerator ash was a free resource that served to cover the landfill and remove the ash. The facility now spends $600,000 per year to bring in cover soil, a valuable resource that would be much better used where it's supposed to be—in the ground.

Over the years, incinerators have become much safer, more efficient, and reliable, making it so that "you take ten garbage trucks in and only one out," as Hawes put it. The benefits to the landfill aren't just free cover ash, but an overall extended landfill life expectancy. The more well-managed and safe a landfill is, generally the more expensive it is; but on the whole, the biggest expense of a landfill is the amount of space it occupies—a resource many places don't have much of, and why many parts of Europe and Asia rely on incinerators as their primary source of disposal. With newer, stricter government regulations for building and operating landfills, in conjunction with growing populations and land becoming increasingly expensive, there is a strong incentive to extend the operational life span of existing landfills. While the ton-over-ton cost of dumping in landfills is cheaper than incinerators, the future cost of closing a city landfill only to build a new one is potentially much greater than the cost-benefit of having a more integrated management system.

The Hanover facility is advanced in their approach to integration, reuse, and sanitation. From an environmental health perspective, when it comes to leachates—all the materials that seep down through the trash and into the groundwater when it precipitates—they got ahead of the curve by implementing protective landfill liners two decades before the EPA mandated all landfills install them. The wastewater created by the leachates is run through a reverse-osmosis process on site that purifies the water to an EPA-mandated level that's cleaner than drinking water before pumping it into the Cape Fear River.

The successful integration of the New Hanover facility comes from establishing a system of material-specific reclamation and recycling programs. As a means of preventing food waste from entering the landfill, they developed a composting program that now produces five tons of compost per week. These programs exist around the world and are becoming increasingly popular in municipal facilities and private composting companies. A construction and demolition program also works to divert construction waste, which makes up about a third of what comes into the landfill, for reuse or recycling. Most of the construction waste is brought in from active construction, but there's an increasing amount of demolition waste, as real estate development runs out of unoccupied space and moves on to knocking down existing structures.

The Burning Question

Incinerators have a social and governmental situation that closely resembles that of nuclear: hotly debated from all sides. Reliance on incineration is often dependent on resources available. In this case, the primary resource is land, making it most popular in countries with limited space or densely populated areas. Lots of people despise incineration as an option for waste disposal, citing air pollution, high costs, heavy metals, and toxic chemicals being released into the atmosphere, water supply, and soil. Incineration is more costly to operate, both to build the facility and to maintain its efficiency as it ages, but it reduces transportation costs and emissions from trucks when incinerators are more urban—a possibility thanks to no smells or noise pollution.

While it's true that incinerators create pollutants, they are generally far less polluting than landfills. The image of an incinerator with toxic smoke (similar to that of a nuclear power plant) isn't a reflection of a modern-day incinerator. They produce particulates, gases, and ash that contain toxins; however, with up-to-date technology, incinerators have the capability to nearly eliminate these toxins from reaching the public. A 2009 study by the EPA found that WTEs produce half the pollutants and nine times the amount of energy as landfills.[2] The volume of waste is significantly reduced compared to landfills, and the leftover ash can be handled by special facilities or even repurposed into construction materials.

Some advocates against incineration believe that the need to "feed" the plants will be encouragement to produce the waste necessary to keep the plants running, rather than encouraging waste to be recycled. Personally,

TRASHY TIDBIT

Tire Fires

1989 in Heyope, Wales: Ten million tires caught fire and smoldered for over fifteen years.[3]

1983 in Frederick County, Virginia: Five to six million tires burned for nine months, sending up a fifty-mile-wide plume of smoke. Eight hundred thousand gallons of low-grade oil were gathered for reuse in a process called pyrolysis, which captures the gases released when tires burn. It was declared a Superfund site.[4]

1984 in Everett, Washington: Four million tires known to locals as "Mount Firestone" caught fire and burned for months, with smoke blocking the sun by the end of the first afternoon. Ironically, the tires had been collected in an old city landfill with the intention of selling them to create fuel.

1999 in Westley, California: Seven million tires caught fire when struck by lightning. The oil excreted by the tires flowed downhill and also caught on fire, with 250,000 gallons of oil recovered from the site.

I think this is a weak argument. Yes, in an ideal world we would have more items that can be recycled and recycling would be effective and efficient, but that's not the current reality. To block improved technology that could help a current crisis does not necessarily promote a resistance to change. It is like saying we will not build electric cars because it will discourage the country from switching entirely to biking. We will continue to have extraordinary amounts of trash for a long time, there is no doubt about that, so methods of efficient disposal need to be implemented sooner rather than later. The European countries who have the highest use of incinerators are largely the ones with the highest recycling rates!

In Japan and Europe, the attitude tends to be quite different than that of the United States, with much of the government and public opinion being pro-incineration. Due to a limited amount of space available to landfill waste, burning is very common as the primary method of disposal, and in turn, as production of usable energy. In 2017, Denmark opened its most high-tech incinerator (built with a functional ski slope on top) that claims to have almost net-zero greenhouse gas emissions and a high output of energy. The plant generates heat for seventy-two thousand homes in the surrounding area.[5]

Incineration is not the most efficient source of energy possible, but the production of energy while disposing of waste is more positive than generating nothing but toxins.

Trashy Time Capsules and Gas Traps

Landfills are indicators of our global socioeconomic patterns and evolving politics. A barometer of economic booms and crashes, "the trash man is the first one to know about a recession because we see it first."[6] Dumps around the world look and feel very different depending on the wealth and regulations of the region, but all of them have layers of history piled up and continue to rapidly fill or reach capacity as we continue to generate waste. If there is space, we'll fill it. They are evidence of the shifts in what we throw away over time, serving as a time capsule of sorts.

Landfills in the United States became more formalized with the opening of Fresno Sanitary Landfill (FSL) in 1937, which became a model that most modern landfills would follow—using trenches, layering dirt and garbage, and compacting the layers.[7] The site was designated as a National Historic Landmark after it closed in 1989. Soon after the opening of FSL, official landfills started cropping up around the country, with Staten Island's Fresh Kills Landfill opening its site in 1948. By 1955, it was the largest dump in the world, spanning an area three times the size of Central Park, filled with a twenty-story mound of 150 million tons of garbage.[8]

The landfill closed in March 2001 but was quickly and unfortunately revitalized that September. The debris from the September 11 World Trade Center attack was taken to the Fresh Kills landfill, where it was further searched through for victims and other artifacts related to the event. During the cleanup and disposal process of 1.8 million tons of debris, the landfill became a site of investigation and recovery. Operating twenty-four hours a day, workers separated materials and inspected evidence; eventually, as the weather turned colder, the Army Corps of Engineers constructed greenhouse structures on the property to continue the process of sorting.[9] The investigation became the most costly in US history: $1.5 billion to clean up.[10] After nearly a year, the landfill closed for a final time in July 2002, ending on a dark, emotional note for New Yorkers.

While most landfills aren't the site of tragedy or investigations, Fresh Kills and other huge landfills have taken a similar trajectory, undergoing a transformation from open pits of garbage to usable recreation areas. Former landfills throughout the country have become public parks, occupying an estimated 4,500 acres in the United States. From Red Rock Canyon Open Space in Colorado to Millennium Park in Boston to Mount Trashmore Park in Virginia Beach, landfill parks are everywhere. These former dumps are hosts to major events both past and present, such as the annual hot air balloon festival in Albuquerque, New Mexico; the World's Fair (twice) in Flushing Meadows Corona Park, New York; and the international kite festival in César Chávez Park in Berkeley, California. If you've flown into John F. Kennedy Airport or been to a Chicago Bears game, you've been on a former landfill.

Creating a park from a landfill has its advantages and challenges. Closed landfills are generally very close to urban areas, tend to occupy large open spaces, and are usually cheap or free for cities to purchase due to their uninhabitable state upon closing. If you've ever flown over Los Angeles, you've seen its smoggy sprawl, so it's not surprising that one of the largest landfills in the world lies outside the city, occupying 630 acres and rising five hundred feet. Operational from 1957 to 2013, the Puente Hills Landfill collected more than 130 million tons from the Los Angeles area. The magnitude of the landfill was so great that it created its own microclimate, requiring fans on its border to keep the winds from blowing into the surrounding communities.

Because closed landfills need massive transformations to become usable, they can be incredibly challenging to turn into park land due to severely contaminated and toxic materials, unpredictable settling, and gases emitted from the decaying material. Methane begins forming when garbage starts to decompose without oxygen. Currently contributing to 11 percent of global methane emissions, municipal solid waste landfills are the third-largest source of methane production, and that's slated to rise significantly with increased global trash production. Landfill gas is made up of a combination of methane and carbon dioxide (both greenhouse gases), and while CO_2 is our poster child for climate change, methane is twenty-eight to thirty-six times more potent as an agent of global warming, trapping far more heat than CO_2. The scale of landfill gas emission is huge, and the potential to capture and use the gas would result in production of a highly valuable resource—turning something incredibly dangerous into a renewable energy resource to offset coal and gas use.

The EPA now closely monitors the building and maintenance of operational landfills, overseeing the planning and implementation of landfill parks. It's imperative to create spaces that do not pose threats to the health of park users, and devise systems to either cap or use the methane gas released. Some retired landfills' methane gas provides energy to surrounding areas, often in high enough amounts to offset costs of maintaining the landfill, in addition to selling energy to the local grid.

The Center for City Park Excellence estimates the cost of converting a landfill into a park to be $300,000 per acre. By that estimation, a landfill the size of Puente amounts to $189 million to convert it into land usable by the public. Despite the extensive planning time and cost involved to create these parks, they are a helpful way to return land trashed by landfills into green spaces that can provide habitat for wildlife and usable outdoor space for urban areas.

There is no cut-and-dried answer to the landfill versus incinerator debate. There are benefits and problems with both methods, and neither is perfect by any means. If they're built and operated well, and if measures are in place to ensure more efficient, safe standards are met, both can serve as decent options for now. Ultimately, the amount of waste we produce needs to severely decrease to allow both types of facilities to have longer life spans and less impact on the environment.

Part 3

WASTE
AT HOME

Chapter 7

FOOD

ood waste is the biggest sector of municipal solid waste (MSW) in the United States. I have a hunch that will come as a surprise to many, because what we hear about most is plastic. Plastic is the buzzy big problem in the garbage world due to its indestructibility and visibility. Food waste, while a huge problem, has much more tangible solutions. It has a direct use both before it is thrown out (feeding people) and after (feeding animals and composting), which provides more potential ways to solve or reduce the severity of the issue. Reduction of food waste can happen through multiple channels: redistributing food from those with too much to those who cannot access enough; providing education around expiration dates and enforcing rebranding from marketers; and an overall restructuring of food production, which of all the facets is certainly the most complicated and grand.

According to the World Bank, 44 percent of global waste is organic: edible food and green matter such as yard waste.[1] In the United States, an estimated 50 percent of produce is thrown away, and food waste as a whole accounts for 33 percent of the total MSW—the largest portion of our total household garbage.[2] The environmental toll is high, as is personal financial loss. The average US household of four throws away between $1,365 and $2,275 of food each year.[3] That number is appalling in large part because almost all organic materials that are thrown away can be composted, fed to animals, or given to those in need.

Edible food waste is a relatively new problem. Until the Industrial Revolution and the implementation of municipal waste collection systems, food was not grown or purchased in excess; scraps were fed to livestock or were the base of broths or other dishes that could use bones and limp veg-

"TRASH FISH"

"Trash fish" (also known as "rough fish") are fish that are considered monetarily worthless, and there are often no ecological protections or limits on fishing and/or dumping carcasses of unwanted trash fish. Which species are considered valuable varies between locations and cultures, but generally having some species deemed worthless ends up harming their populations and the surrounding ecosystems.

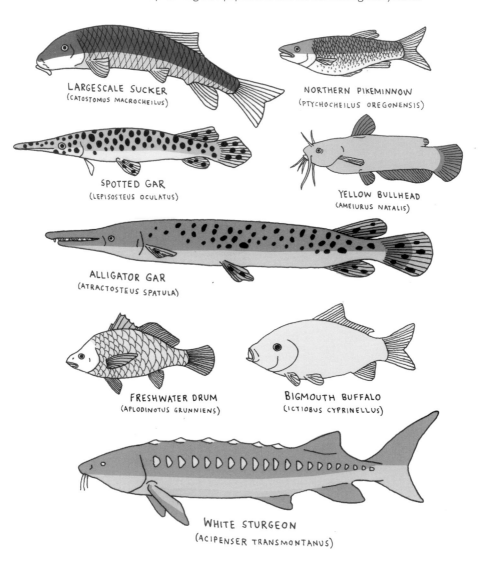

LARGESCALE SUCKER
(CATOSTOMUS MACROCHEILUS)

NORTHERN PIKEMINNOW
(PTYCHOCHEILUS OREGONENSIS)

SPOTTED GAR
(LEPISOSTEUS OCULATUS)

YELLOW BULLHEAD
(AMEIURUS NATALIS)

ALLIGATOR GAR
(ATRACTOSTEUS SPATULA)

FRESHWATER DRUM
(APLODINOTUS GRUNNIENS)

BIGMOUTH BUFFALO
(ICTIOBUS CYPRINELLUS)

WHITE STURGEON
(ACIPENSER TRANSMONTANUS)

etables. There were no expectations or beauty standards applied to produce, and *edible* was edible. Anything rotten, as well as peels or scraps, were tossed out the kitchen window to decompose or feed to animals. Pigs, goats, horses, probably raccoons, and chickens were there to consume the food scraps, manufacture fertilizer to grow more food, and eventually be food themselves. There was no incentive to grow more than necessary, as farming requires labor, water, and land that are wasted if not everything grown is consumed. Paradoxically, while a third of the food in the United States is thrown away, a sixth of Americans are food insecure.[4]

Almost all food scrap material could be diverted and actually be beneficial to the planet with concerted effort and infrastructure in cities, which is starting to be more common. Home, private, and municipal composting systems increase the health of soils and reduce the contribution to landfills. Composting is a step toward a more circular system of reusing those materials in a way that is ultimately cost efficient. Reuse doesn't only mean bringing a cloth bag to the grocery store. Replenishing soil health with food scraps can be a form of reuse—getting more life from a material that would otherwise be thrown away.

Water Waste

The reason behind imbalanced food waste is, like most global systems, complicated and huge. Waste happens at all stages of the process, from growing to consumption. Starting at the growing process, we have water waste. Globally, agriculture uses 70 percent of our fresh water, a finite, precious resource, so water waste is a big side effect of food waste. With such a huge percentage of food being wasted, that same amount of water goes with it. It's a good reminder that no garbage is isolated to the object being thrown away; there are hidden costs to everything. There is waste before anything even gets to the landfill.

The amount of water waste in food production is astonishing. The world is experiencing a water shortage crisis, with five billion people expected to experience water scarcity by 2050. Agriculture uses significant amounts of groundwater that is tapped to irrigate crops, and in places like California's Central Valley, which is the US's top agricultural hub, they're quickly running out of plentiful usable water. Unsustainable groundwater irrigation has contributed to reservoirs drying up or depleting at a rate that replenishment can't catch up to. In response, farmers are cutting back on the amount of crop they're growing, and selling water to places that face even greater scarcity, making money without the risks of farming.

In places where the groundwater has been severely diminished, farmers are making desperate attempts to encourage groundwater replenishment. Many crops have high water demand, which leads to depleted water sources and smaller and smaller harvests. However, there are farmers in the Southwest employing tried and true methods of low-water-use growing. These techniques have been around forever, and follow the ecosystem's natural rhythms and mechanisms to benefit crop yields and diminish resource demand. Worms, compost, and intentional layered planting all aid the health of soil, which in turn aids the health of its crops. The richer the soil, the more water it holds. If you think about watering a houseplant whose soil has become bone-dry, you know that the water runs straight through it, not saturating the soil or plant's roots at all. But soil that is rich and moist holds the water well, sustaining the roots for longer with less water, leading to a happier plant.

The Hopi and Tohono O'odham Nations in Arizona practiced these methods in dry climates long before large-scale, water-intensive, irrigation-driven agriculture started there. Growing drought-resistant crops suited for the harsh environment, intercropping, and capturing rainwater creates a stewardship of the land, rather than an exploitation of its resources. While some small-scale farms are learning from those who have been tending the land for millennia in the United States, most are not. The majority of farms use huge amounts of resources and labor, only to leave rotten or aesthetically imperfect produce in the fields, even if they are perfectly edible but not perfect looking.

Pretty Produce

The idea of aesthetic perfection started to grip American households and supermarkets in the 1940s and 1950s, with produce being no exception. As families had more disposable income post-WWII, indoor home refrigeration combined with increasing accessibility to internationally imported produce created the expectation of picture-perfect food to match the idea of a picture-perfect household. Some farms currently have contractual obligations and expectations to provide produce of a certain aesthetic caliber to satisfy their distributors. But when they've nailed it with a high yield of "perfect" crops, an overabundance of desirable fruits or vegetables can cause a market surplus, and in line with basic supply and demand economics, can drive the price way down.[5] This puts farmers in a bind: when there aren't enough flawless tomatoes in a harvest, farmers still leave the underripe, overripe, ugly, or damaged ones unharvested in the field. But when a harvest produces almost all desirable tomatoes, there isn't enough demand in the market to sell them all.

Historically, there has been a natural relationship between overproduction and feeding those in need. Gleaners (people who cull unwanted produce from fields) have long been a fixture of food recovery. Featured

TRASHY TIDBIT

March 5, 1973:
Great Michigan Pizza Funeral, Ossineke, Michigan

Mourners watched as 29,188 frozen mushroom pizzas (still in their wrappers) were dumped into an eighteen-foot-deep hole. Deemed unsafe by the FDA due to potential botulism found in the canned mushrooms, the pizzas were recalled and ceremoniously buried, topped with a white-and-red wreath to symbolize their cheese and sauce. In attendance were hundreds of townspeople and Michigan's governor, who gave a speech. Pizza was served at the funeral.

in the famous painting *The Gleaners* by Jean-François Millet in 1857, all the way back to the Old Testament, gleaning is a method for feeding those who can't afford to purchase enough food. In eighteenth-century Europe, gleaning was a legal right for those who didn't own land, for widows, and for people with disabilities. The practice alleviated food insecurity and therefore waste, encouraging access to all.

The Gleaners (Des glaneuses) by Jean-François Millet

The two opposing problems of excessive food waste and food insecurity seem as if they should function to alleviate one another if not solve their paradoxical crises; it seems like such a straightforward, mutually beneficial solution, so what are the obstacles preventing a widespread effort to reduce the gap? With increasing awareness of food waste issues, gleaning is resurfacing in an organized way through volunteer community and nonprofit groups focusing on food justice to mitigate food insecurity. The harvesting programs in large part depend on volunteers, which is an unpredictable source of labor. To make matters a bit trickier, the moments that are ideal for collecting unwanted food can be unpredictable as well—

sudden heat waves, floods, or droughts can affect harvesting operations and quickly turn the quality of produce.

In the grim picture of harvesting waste, there's a small sliver of good news, which is that the unpicked produce is generally left in the field, tilled back into the soil, and used as compost, rather than being sent to a landfill like household, supermarket, or restaurant food waste.

Homegrown or Exotic

Once picked, most produce is transported, often very long distances. California is the largest domestic growing state, but more than a third of all vegetables and over half of fruits purchased in the US are grown internationally, mostly in Central and South America.[6] Transport is energy intensive and often involves pesticides, bulky packaging, and refrigeration to preserve the produce for its potentially long journey to supermarkets around the world. Once it arrives, it can sit in refrigeration for months to sustain supermarkets' supply through the off-season. But the cost-benefit analysis of domestic versus imported food isn't as simple as "far away is unethical" and "nearby is ethical" when it comes to wastefulness.

As international migration expanded over the last hundreds of years, by force or choice, people brought with them foods from home. Some of these previously exotic foods have become mainstream. Enslaved people from Africa brought peanuts, black-eyed peas, okra, and hot peppers to the United States, and colonists in the tropics took bananas and pineapples back with them. Colonizing Europeans were obsessed with the pineapple, describing it as "the prince of all fruits," going to great lengths to grow it in colder climates, and devoting extra resources to cultivating it, making it a status symbol of wealth.[7] Nowadays, if you see one upside down on a porch, it's a subtle invite to a swinger party inside.

While pineapple used to be an imported good reserved for the rich, nowadays importing is often the cheaper option for year-round produce. Considering that large-scale food production is very complex no matter how you slice it, off-season imported produce raises the question of domestic versus international produce and the cost-benefit of both. Is it better for an in-season tomato to be shipped overseas fresh from the vine than grown off-season in a hot house? Other counties might grow in ways that produce much less waste on a farm three thousand miles away. By benefit of the location, that farm uses less irrigation for crops that are naturally suited for the local environment, fewer mechanized tools, and more natural fertilizers than a farm that's four miles from you.

Sure, it would be great if seasonal produce, locally sourced meat and dairy, and fresh-baked bread were accessible and affordable for everyone. In many places around the world, that localized way of growing and selling food is the available and affordable option. But like many things, the curve can be wonky. I very occasionally go to the farmers market, but generally avoid it because of the social hell of running into anyone and everyone from my entire life there. Even so, I like supporting local farms and growers, bringing home less packaging, and being motivated to eat the nice fresh carrots rather than passively letting them become a limp mess. But farmers markets alone don't solve the problem. The United States isn't oriented around affordable local markets like much of the world is. Many countries operate in a more balanced way. They have permanent year-round local markets that operate daily and offer a wide range of foods, from bulk spices to fresh fish to fruits and vegetables; but they also have chain supermarkets with imported packaged foods. A general trend tends to correlate increasing income with packaged food consumption in places undergoing a transition to more disposable income, women entering the workforce, and an increase in eating outside the home.

Food Waste Happens at Home

Once food leaves the fields, the majority of waste happens at home, accounting for 40 to 50 percent of all food waste. A major hurdle in reducing food waste is addressing misleading information about expiration dates—info that I'm sure has confused all of us at some point or another. I had the pleasure of working with Mei Li and Irene Li, founders of Food Waste Feast, on their 2023 cookbook, *Perfectly Good Food*, which focuses on reducing food waste through creativity and education. They explain that expiration dates are purely opinions about a product's peak freshness, whether that's indicating crunchiness or most flavor, but not about its safety. In most cases, take expiration dates as a suggestion and not an order, as there is little to no regulation or standardization for food manufacturers to adhere to when determining expiration dates. There are obvious signs for when most food is bad: texture, appearance, and smell are pretty good indicators that something might be off.

In addition to being aware of misleading expiration dates, limiting food waste food at home can be reduced by being more aware of what you already have. There is most likely something on the verge of being limp or stale at all times, but those things are often still salvageable. My constant limp food is celery. There seems to always be a half-used, pathetically limp bunch in the back of the fridge. For soup it doesn't matter, but for dishes where its only purpose is to be crunchy, it does. Mei and Irene give a great tip that is handy for celery but also applicable to a lot of perishable foods that you anticipate will get lost in the back of the fridge: give them their own zone front and center. When going shopping, meal planning, or searching for something in a crowded fridge, having a visible area where it's clear what isn't necessary to buy more of or what might be good ingredients to plan meals around can help save those things from going into the trash. If I were to actually use the "eat me first box," as the Lis call it, and make soup or stock to freeze, I would probably save all the celery I buy.

Restaurant Waste

Restaurants aren't quite as bad at wasting, making up 10 percent of food waste, which reflects the economic ramifications of overbuying and/or overserving. The financial loss of food wasted at home is a little less obvious, as it's on a smaller scale and over a longer period of time. Catering is by far the worst waste generator of the prepared food world. The Lis estimate that half of plated, catered dinners are wasted with no provided option to take home leftovers. I've never been to a wedding or event that handed out Tupperware to the guests. Restaurants have an incentive to streamline and plan well to avoid waste as much as possible. This obviously has exceptions, but on the whole there is a bottom line that benefits from as little waste as possible, and whatever meal leftovers leave the restaurant are up to the patrons to hopefully not toss.

There are organizations popping up to curb food waste from restaurants, including Too Good To Go, a certified B Corp that serves as a platform for restaurants to offer discounted meals that would otherwise go to waste. Available in seventeen countries as of 2023, Too Good To Go helped restaurants save 52,554,009 meals from being tossed in 2021. They also implemented a "look, smell, taste—don't waste" label in several countries that encourages consumers to self-assess the quality of a product rather than rely on the expiration dates alone.[8] Humans have been determining whether food is good or bad with our senses for all of time, and all animals (except dogs) have a pretty innate sense of when something is safe or dangerous to consume.

Food Packaging

The actual food and resources wasted make up a huge portion of the food waste issue, but packaging is another piece of the equation. From landfills to beaches and roadside litter, food packaging is extremely prolific in its visibility. We've all likely seen stretches of highway covered with an enormous number of Styrofoam cups, fast-food wrappers, and soda cans. It's one of the sectors of waste that, as individuals, we handle the most frequently.

Plastic bags from bulk produce, plastic packaging of processed food, waxed cardboard cartons, and takeout containers are tossed all the time in most households. You may have seen stores selling two cucumbers or potatoes Saran-Wrapped to a Styrofoam tray—something that has always been extremely bizarre to me. It not only adds physical waste with unnecessary packaging, but forces consumers to buy a predetermined amount of perishable food that might be beyond what the consumer needs. I hate cucumbers, and if I am being forced to use one in a recipe, I certainly don't want another one haunting me from my fridge.

Manufacturers now rely on packaging not just for transportation but for advertising. Removing their branding from their product for the sake of being less trashy is a threatening prospect. When you go to the co-op and crunch the peanuts into peanut butter in that machine, do you have any idea what company those peanuts are from? Probably not. Buying bulk goods means a huge reduction of waste by allowing shoppers to bring their own reusable containers to fill, but going package-free means producers must be willing to sacrifice their constant advertisements every time you walk down the grocery aisles. There can never truly be no packaging.

Things have to get from one place to another somehow, otherwise we'd all be visiting the field or dairy farm to go grocery shopping. But when possible, buying bulk is a great, affordable option, with many smaller operations and some larger chain grocery stores offering a wider variety of bulk goods, including nonfood items like shampoo, laundry detergent, and spices.

Food choice is something inherently overwhelming. There are too many options and too many factors that go into every decision, from brands to environmental claims to expiration dates to even figuring out which store to go to in the first place. But of all the garbage issues to come, food waste is one area that consumers really can improve upon in their own homes. Packaging, growing, and transportation all need an overhaul on the larger level, but being aware and educated on how best to keep your food out of the trash, both by eating it before it spoils and having a place to put it that's not into the municipal garbage bins, is important.

I live alone and I'm picky, so I do end up wasting some food, but I now stick to buying the same fifteen boring foods that I'm guaranteed to eat. I've been trying to use my freezer more when I have a sense something I make today is not going to be appealing two days from now, rather than realizing that after not eating it for three days. Or I freeze things I buy in bulk when my eyes are bigger than my stomach (thanks, Costco). Donating food to local food banks or nonprofits that are working to improve food access in your community is also a great way to extend the life of your unwanted but still edible food. Some stores offer donated excess food at the end of the day to pick up and bring to food pantries or community meal centers. If you have the privilege of consistently abundant food access, practice being mindful of what comes into your home and what goes into the garbage.

Chapter 8

PLASTIC

Think through your day and try to imagine each plastic item you might have encountered.

Within the first twenty minutes of being awake, it's likely that the plastic count is already in the dozens. Your alarm clock (or phone) goes off and you raise the blinds to peek out at the bird feeders outside the PVC-trimmed windows. Sit up, put on your glasses, and slip on fleece-lined slippers. Flip on the light switch in the bathroom and step on an anti-slip bath mat to brush your teeth, maybe coaxing out the last of the toothpaste from the tube using one of those little rollers. Perhaps a bit of flossing, too, if you're feeling diligent? (I never am.) A quick face wash from a pump bottle, dry off with a polyblend washcloth, and put on a blob of sunscreen for good measure. Pour coffee from a bag into a grinder, measure it out with a little scoop, brew it in an electric coffee maker, and top with half-and-half (after pulling off the little plastic seal ring). Cook eggs from a protective carton in a nonstick skillet. Peel the little sticker off an orange while reading the paper that arrived in a bag to keep it from getting wet in the rain. Scrub the plate with a sponge and dish soap and leave it to dry on the drying rack.

After using twenty-eight plastic items before getting dressed, you still have twenty-three hours and thirty minutes left in the day to interact with the plastics of the world.

Dozens or hundreds of plastic materials surround us, so integrated into our lives and so well camouflaged that most of the time we're not taking stock of how prolific they are. I see over fifty plastic-based objects from the couch I'm sitting on (propped up with polyurethane-stuffed pillows). The planter in the corner is under a hanging light with a plastic cord; my knickknack collection is on a shelf colored with spray paint (which is basically spray-on plastic);[1] my stack of board games is next to a stack of DVDs. The mural on the opposite wall from me was done with latex paint, and fiberglass insulation fills the walls just behind it. I'm snacking on chocolate chips from a metallic-y plastic bag, and potato chips from a crinkly one. The vacuum cleaner in the corner is silently shaming me for not being used this week, and it's next to the collection of small plastic sharks on the bookcase. My dog's collar is a nylon blend with a plastic snap, and his microchip may very well be made of plastic. This computer, my phone, a tape measure, a remote, a loose light bulb, an empty Tupperware container, a hardcover book, and a tennis ball can. I'm a messy person, and that mess has a whole lot of plastic as part of it. I would imagine my plastic count is most likely less than the average household in the United States due to the fact that I have no children and live alone—it just happens to all be on the coffee table right in front of me.

Plastic was a groundbreaking wonder substance when first developed in the mid-1870s. Named for its malleable nature, plastic would come to change the planet forever—the way we live every day and the way we will leave the planet once we're gone. It's a phenomenal material, able to take on just about any shape, hardness, color, translucency, size, or texture, and at a very low cost to produce, which means it's often capable of replacing natural materials. Plastic's reign emerged from the desire to use it as a substitute for existing materials, imitate the properties of those materials, or push innovation further with the aid of a new substance unlike any other.[2] In order to be successful as a material, plastic needed to serve as a utilitarian alternative to resources that were more expensive to extract or labor-intensive to produce, including wood, ivory, and metal.

On a global scale, plastic has fallen from grace as a novel miracle material and is now the poster child of modern trash production. Campaigns about recycling are oriented around plastics, single-use grocery bags are noticeably gone from some cities, straws are enemy number one, and images of dolphins with bottle rings caught around their noses have been tugging at our moral heartstrings for years. With 91 percent of the world's total plastic mass landfilled, dumped in the ocean, or incinerated since its inception, there's a good reason why plastic is probably the first material that comes to mind when you think about trash.

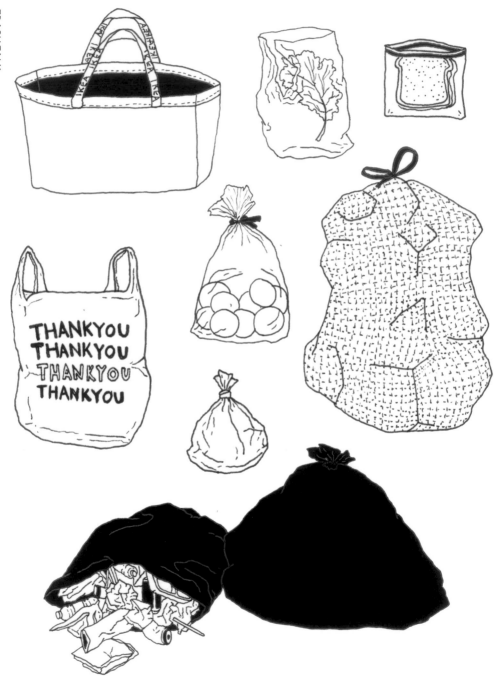

What Even Is Plastic?

Plastic is a catchall name for a huge class of various materials that all use polymers as their main ingredient. I hadn't really thought about polymers much since high school chemistry, and I'm pretty sure I didn't really understand it very well at the time. I won't force you to skip this paragraph by explaining in depth what a polymer is, and will instead summarize the situation: Most plastics are derived from fossil fuels—oil, natural gas, and sometimes coal. The oil is extracted and refined before undergoing polymerization: a process in which monomers form long, chainlike molecules known as polymers (don't worry about it). Once that's done, the various polymers and other materials (bear with me) are blended to form the specific type of plastic material desired—a process called compounding. The numbers on the bottom of containers indicate what their chemical makeup is like—for example, PVC (3) or HDPE (2). Each of these has a different level of recyclability, and municipalities accept and reject certain numbers depending on their facility capabilities and profit margins. This inconsistency does not make for simple rule adherence by residents attempting to correctly sort.

There is increasingly more money, public attention, and effort being put into the development of plastic alternatives, or "bioplastics," that are not derived solely, or at all, from fossil fuels. The term is a little misleading, requiring products to be made from only 20 percent or greater renewable materials. Bioplastics are quickly growing to meet corporate pressure to greenwash products; there are many benefits to using bioplastic instead of petroleum-based plastic when it's produced and disposed of correctly. Though it needs to be said: corporate advertising is not known for its honesty, and when it comes to bioplastics, they often mislead customers, leading to more problems than good much of the time. Many companies are trying to balance their bottom lines while saving face in the growing environmental movement, and this has created some deeply confusing messaging around new materials labeled "biodegradable" and "compostable." And that's not even accounting for the long history of issues with the classic recycling label (see p. 87).

Biodegradable and compostable goods seem like good ideas, right? A step in the right direction? It's important when thinking about solutions to trash problems, particularly plastics, to acknowledge that there isn't

TRASHY TIDBIT
Mr. Trash Wheel and Friends[3]

In the Baltimore Inner Harbor resides a family of semi-autonomous trash-eating
water wheels. Invented by John Kellett, Mr. Trash Wheel has been collecting
trash since 2014. The family has expanded since then to include Professor Trash
Wheel, Captain Trash Wheel, and Gwynnda the Good Wheel of the West.
As garbage empties from the city's storm drains into the harbor, it's removed
by solar and hydropower conveyor belts into a dumpster. Gwynnda is the
largest of the group, raking in an estimated 300 tons of trash annually.

Since its installation (as of 2023), the family has collected a combined:

938,626 plastic bags
13,036,008 cigarette butts
1,812,576 plastic bottles
1,383,053 Styrofoam containers
1 snake

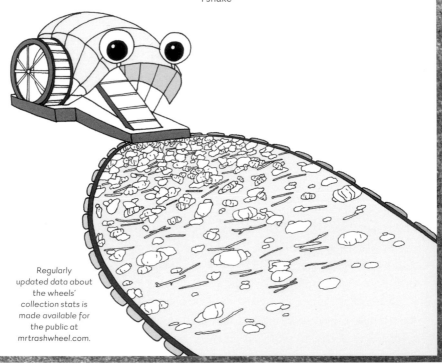

*Regularly
updated data about
the wheels'
collection stats is
made available for
the public at
mrtrashwheel.com.*

a silver bullet in terms of innovation. In theory, these materials should be able to break down, solving the issue of persistent petroleum plastics. But in reality, they're unreliable labels that can't yet fit into our existing disposal systems. The ways that garbage is disposed of in a landfill greatly vary in temperature, oxygen levels, and sunlight—all of which affect the speed and effectiveness of plastic degrading. Recycling has been so deeply ingrained as the way to solve the problem that recycling a bioplastic feels like doubly the right thing to do: win-win. But one irony of all of this is that, more frequently than not, biodegradable plastics can contaminate standard recycling operations when mixed with petroleum plastics. When added to landfills, they can release the greenhouse gas methane over centuries because of their organic compounds.

In another experiment of plastic substitutes, if I were to give you a cup with an official "compostable" label on the bottom, would you feel comfortable, or dare I say, confident, to toss it in with your eggshells and apple cores in your backyard? According to the EPA, in order to be labeled as "compostable," it must decompose at a rate similar to or the same as the other organic matter being composted. What's misleading about materials advertised as compostable is that these products are intended to be composted at industrial scales—not in your backyard. Most households and public spaces do not have compost bins or pick-up services, so when compostables end up in the landfill instead, they don't become soil-enriching material as intended. Unless it says clearly, "home compostable," on the label, assume it's not.

The Coca-Cola Problem

With an average of $4 billion spent on global advertising per year, Coca-Cola has established itself as the most recognizable beverage brand in the world, and in turn makes them the greatest source of single-use plastic bottles.[4] As of 2022, Coca-Cola produces over one hundred billion single-use plastic bottles a year, earning them the top spot on Break Free from Plastic's "world's worst polluters." PepsiCo, Unilever, and Nestlé trail behind, with less than half the amount of waste as Coke.[5]

Coca-Cola's history of making and breaking pledges to improve their mark on the environment is long and ostentatious. Promises in the 1990s to reduce virgin materials by increasing recyclates in their plastics have not been met, yet they continue to make these pledges with ever-extending timelines and bigger ambitions. Because they are self-created and self-enforced commitments, there is no reprimand for failing to meet those goals, and realistically, no consumer is paying enough attention to incite accountability through public demand. Corporations have established norms that need to be overhauled, and having a different form of single-use plastic bottle isn't getting to the root of the issue—our whole way of engaging with use and waste needs to shift.

Some plastics are extremely obvious contributors to mountains of waste—food packaging, shipping materials, bottles, and to-go containers to name a few. We buy, handle, and toss these items daily, taking part in their journey from the store to our bins. But some polluting plastics are nearly invisible, shedding microplastics while in use or once they're tossed, with much of that swirling around in our oceans to the tune of more than 5.25 trillion pieces of plastic on the surface alone.[6] Yikes.

Ocean Plastic

The world has a history of using the ocean as a vast garbage can. The National Academy of Sciences estimated that 100 million tons of petroleum products, one million tons of heavy metals, 4.5 million tons of industrial waste, 4.5 million tons of sewage sludge, and half a million tons of construction debris were dumped into the ocean (just in the United States) every year, along with eighty-nine thousand containers of radioactive waste over the course of 1946 to 1970.[7]

The Great Pacific Garbage Patch has become a popular concept of what ocean pollution looks like: a huge swirling pile of trash on the surface of the ocean. The myth that it can be seen from outer space persists, and while it is enormous and it is trash, it's almost invisible. It's one of five enormous patches, making up a fraction of what's present in marine habitats around the world, most of which sinks into the deep sea. It's estimated that for every square kilometer of deep sea, there are four billion plastic microfibers, with 70 percent of the marine debris sinking beneath the surface.[8] A non-degraded plastic bag was found thirty-six thousand feet below the surface in the Mariana Trench, the world's deepest ocean ravine.[9]

TRASHY TIDBIT

1992: Friendly Floatees Spill, North Pacific Ocean

A container from a cargo ship carrying plastic bath toys from China to the US fell off the ship, spilling twenty–eight thousand toys, many of which were rubber duck–ies, into the Pacific.

Each year, fourteen million tons of plastic makes its way into the ocean from boats, sewage or storm water runoff, beach litter, and rivers.[10] The majority of ocean debris is microplastic—plastic particles measuring less than five millimeters in size, which is about the width of the eraser on a no. 2 pencil. Microplastics are either broken down bits of larger plastic items that have degraded over time or pieces, known as microbeads, that are intentionally small and used in manufacturing everyday products, such as exfoliating cosmetics, toothpaste, and prescription drugs. Sunlight, turbulence, and oxidation cause plastics to degrade and eventually disperse into thousands or millions of particles, which are carried along currents to end up around the world. Larger items such as single-use plastic bags (*five trillion* per year) and microplastics alike can end up thousands of miles from their point of origin.

Sailing the Plastic Seas

To highlight the journey plastics can take around the world's oceans, from March 20 to July 26, 2010, a boat made from reused plastic products sailed from San Francisco, California, to Sydney, Australia. The brainchild of Michael Pawlyn and David de Rothschild, *Plastiki* was based on the concept of *cradle-to-cradle*: continuous reuse of products in a closed loop that mimics biological processes, such as nutrient cycling.

Plastic bag thirty-six thousand
feet below the ocean surface
in the Mariana Trench

36,201 ft.

Mount Everest has become the world's highest garbage dump, with tons of waste left behind by climbers who ascend the mountain. The Mountain Clean-Up Campaign conducts cleanups with the Nepal Army, bringing back thirty-four metric tons in 2022—an amount that increases with every year as accessing the mountain becomes easier for the wealthy. Currently, climbers are required to pay a refundable $4,000 deposit to the Nepal government in exchange for bringing back eighteen pounds of garbage. For context, the average trek up Everest costs, as of 2023, around $50,000, with some people paying more than $100,000 to summit the mountain.

The boat was composed of 12,500 plastic bottles and other PET plastic products, held together with naturally derived adhesives, and powered, along with sails, by renewable energy sources, such as solar panels and wind turbines. The idea behind the journey began after de Rothschild learned about the extent of plastics in the world's oceans, a large part of which is caused by the deterioration of these same products. De Rothschild states, "Bottled water epitomizes the absurdity of our throwaway society. Each and every day, Americans consume 70 million bottles of water—nearly 9 billion gallons of bottled water a year."[11]

The project set out to be a model for how plastic waste can be a resource beyond single use. It is a physical depiction of one-to-one plastic reuse, but it's primarily a call to action to address the growing problems around the environmental impact of garbage. While making boats out of plastic bottles isn't the way global garbage disposal issues will be solved, it is a powerful message about the lost potential of what we consider trash.

Wildlife in Trouble

Large marine animals, many of which are endangered or threatened, such as sea turtles, manatees, whales, dolphins, and seals, commonly eat plastics, with an estimated one hundred thousand animals dying each year as a direct result of marine pollution.[12] Plastics of all sizes affect marine life differently. The largest contributor to marine pollution is abandoned fishing gear, known as "ghost gear," which can cause entanglement or disruptions to feeding behavior. If you imagine a white or clear single-use plastic bag floating in water, being slowly carried by a current, it's pretty easy to see how it could be confused for a jellyfish—a primary food source for some ocean mammals. And smaller plastics can be mistaken for algae or plankton—food for animals higher up on the food chain. But it's not just visually deceptive; experts have speculated that the echolocation of deep-sea divers, such as whales, is disrupted by the particles, perceiving the debris as food. On the surface, seabirds eat floating plastics, often feeding it to their chicks, resulting in their deaths. Since many sea creatures sink to the bottom or are eaten before they wash ashore, it's hard to quantify the true toll of pollution on their health, but it's clear that consumption of debris is damaging to the health of marine animals.[13]

TRASHY TIDBIT

February 13, 1997:
Great LEGO Spill, sea off Cornwall, England

Tiny divers, flippers, daisies, and dragons all fell into the sea as part of a cargo wreck that dropped 4.7 million LEGO pieces into the sea. Unlike the rubber duckies (see p. 119), which have been found all over the world, the collected LEGOs have remained relatively local to the shores of England.

Much like the famously heartbreaking Sarah McLachlan ASPCA commercial from 2007, featuring her song "Angel," plastics such as fishing nets and plastic bags garner our attention through heart-wrenching images of whales caught in fishing line, or birds with plastics filling their stomachs. We relate to animals as an extension of ourselves; witnessing the lovable and relatable lives of whales or dolphins cut short due to direct interaction with our garbage elicits a different response than considering invisible microbeads meandering around the deep sea. The imagery is effective, but it still hasn't spurred big changes. The effects of that type of plastic pollution—large and/or intact gear—are relatively well-known. But microplastics cause a potentially more threatening (and mysterious) problem. It's hard to make a cute commercial about microplastics. Because they can be nearly invisible and much of the oceans have yet to be studied (too big! moving too much!), the ability to measure them accurately and completely is a huge challenge. It's nearly impossible to survey the over 332,519,000 cubic miles of water on the planet.

We are seeing animals interact with small ocean plastics, but not in the way we'd expect. The invisible, enormous Great Pacific Garbage Patch is becoming its own ecosystem. According to a study published in 2023, there are now species ranging from jellyfish to crabs to anemones to sponges all

Between five hundred thousand and
one million tons of fishing gear are
abandoned in the ocean each year.

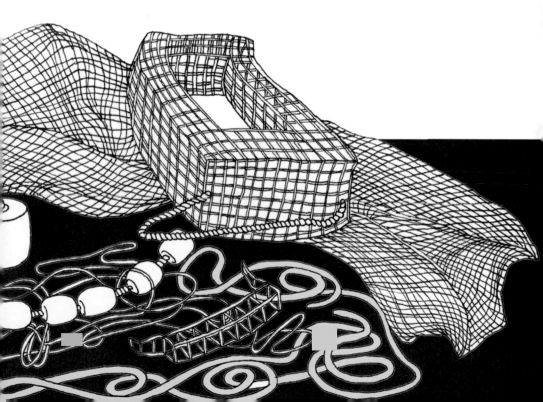

living on the plasticky surface of the open sea, also deemed the "plastisphere." Upon analyzing a teeny sample of the 1.8 trillion pieces of plastic in the garbage patch, researchers found that 70 percent of the identified species in the sample are native to coastal ecosystems.[14] It's hard to imagine a crab living on the surface of the open ocean, and for good reason: it probably shouldn't be there.

Determining the outcome of a bizarro neighborhood of sea life all living side by side on pieces of plastic is hard to predict. Will they stay in the swirl of the Pacific patch? Or float on their plastic to a different destination? Will they breed with one another or compete for food or eat one another? Can they hop from one plastic island to another, or are they bound to their original vessel of transport? Much of the garbage patch is from the 2011 tsunami in Japan, which means that plastics can travel far and stick around, so it's likely that species are traveling throughout the oceans from rivers and coastlines around the world. Approximately 80 percent of the ocean's plastics are from one thousand rivers around the world, meaning that litter, runoff, and storm debris in rivers will likely find their way into the ocean.[15]

The Tire Problem

The problem of seabound plastic, and the hitchhiking creatures along with it, has multiple sources.

Some of this ocean pollution starts on land in unexpected ways. For example, let's take a look at tires. An estimated one billion tires are disposed of annually, heading to landfills, secondary markets, recycling centers, or straight into waterways.[16] In the 1970s, some fishermen with good intentions but a terrible idea dumped two million used tires off the coast of Florida with the goal of creating an artificial reef. Sandwiched between two natural reefs in a part of the world that experiences wild hurricanes that churn up the sea, it's no surprise that these tire mounds, known as the Osborne Reef, dispersed, wreaking havoc to their surroundings. Not only did the tires fail to encourage reef development, but they actively destroyed the existing ones. The project was an utter disaster, and beginning in 2007, cleanup efforts have been in place to remove the tires by both the US military and private companies. As of 2022, a total of 438,545 tires have been recovered.[17]

Car technology has been adapting to reduce waste, reduce emissions, and reduce reliance on fossil fuels, but all road vehicles, regardless of their engine, rely on tires that contribute a shockingly high amount of pollutants to the environment from the continual, imperceptible wear and tear shed by every tire on the road. Cheaply manufactured synthetic rubber has replaced most of the natural rubber in tires, using petroleum and toxic-chemical-laden materials that have no environmental regulations at present.[18] The health and environmental effects of tire pollution are not well documented yet, but localized studies have already shown that there are high quantities of tire particles in marine ecosystems, accounting for an estimated 10 to 28 percent of all ocean plastic pollution.[19] Each commute or road trip casts off more polluting particles per mile than the average tailpipe exhaust, sometimes up to

Trash off the ground: Every year the airline industry produces approximately six million tons of garbage, with each passenger producing 3.15 pounds of waste per flight. The fancier your seat and the longer your flight, the more you'll contribute.

1,850 times more.[20] It makes sense—the roads act like sandpaper, wearing away at our tires for years until they get tossed; that discarded material has to end up somewhere, and that somewhere is oftentimes into the ocean.[21]

So far, there have been no solutions to the plastic-shedding tire problem. However, people are looking at roads and working on ways to make them more environmentally friendly. As a small trial, a Dutch company, CirculinQ (formerly named PlasticRoad), installed a bike path in the Netherlands that is a recycled plastic hybrid material designed to absorb water from heavy rains and release it slowly into the ground, rather than create mass amounts of runoff like traditional pavement. The experiment has proven successfully durable and effective thus far.

On a much bigger scale, India—home to the second-largest road network in the world—has been building sixty thousand miles of partially plastic roads for a decade, with a multifold benefit to the safety of its drivers: reduced infrastructure costs, and a simple method for plastic reuse with no recycling processing needed. The incorporation of plastic has the potential to tackle waste in two ways: keeping plastics out of the landfill or ocean, and reducing the amount of asphalt required for road building. The effort to clean, sort, and recycle plastics is a huge, expensive undertaking, and one that is often relatively unsuccessful, but using discarded plastics in road building doesn't require any of those steps. It can be dirty, mixed material, and easily incorporated into asphalt once shredded and heated. Using these plastics saves them from being incinerated or landfilled and seems to be making the roads more resistant to potholes.[22] Potholes pose a genuine danger to drivers, causing 10 percent of car accident deaths in India.[23]

The still experimental, plastic-infused roadways are installed in Ghana, South Africa, Australia, Vietnam, Mexico, and the Philippines. With seemingly better life spans and a higher resistance to damage due to the elements, plastic roads offer a promising potential for huge amounts of plastic to be diverted from landfills. They also offer low-cost materials for regions that are expanding their paved roadway systems, and act as a landfill diversion in places where there are no recycling systems in place, due to infrastructure or prohibitively high operating costs with low returns.

In addition to the plastic diversion, reducing asphalt use could have a huge impact on reducing construction waste and pollution. We all know that certain hot and sticky asphalt smell; it's extremely recognizable and one that signals a *this probably isn't good for me to be breathing* response. That response is the correct one to have! Making up approximately 45 percent of surfaces in US cities, paved areas are a huge area of opportunity for improvement.[24] Made largely from crude oil and concrete, asphalt is a big waste generator, requiring energy-intensive and waste-producing mining and processing (see chapter 12 for much more on mining).

While this technology has the potential to revolutionize road building and plastic disposal, it's not without its potential flaws. Ten years isn't long in the life span of large-scale infrastructure, and is a short window of time to effectively measure the effects of these materials interacting with the environment. It's a much different scale than in a lab, being exposed to all kinds of environmental elements that can degrade the substance. Asphalt has its negative effects, releasing toxins into the air, and it's unclear whether the incorporated plastics will leach or be released in a harmful way. Plastic waste in the larger systems of oceans, the food chain, and our own bodies are yet to be fully understood, but the possibility of diverting a huge amount of plastic waste while reducing the need for oil extraction could make a big impact. If we can't yet figure out the tire problem, we can at least address the roads they're driving on. And in the meantime, keep your tires fully inflated to reduce particle shedding.

Plastic . . . Eaters?

Innovations around plastic reuse and disposal can be simple and cost-effective. Sometimes nature even makes them for free. There is a group of species with a new name: plastivores. Like herbivores, carnivores, and omnivores, their name alludes to their diet, and in this case, the diet is plastic. This group of organisms has quickly adapted over the course of less than a century to feed off nonbiodegradable synthetic materials. In 2017, Federica Bertocchini, a scientist and beekeeper in Spain, discovered that plastics can be digested by waxworms, the larvae of wax moths—which are despised by beekeepers for destroying honeycomb in bee colonies.[25] Bertocchini gathered the larvae into a plastic bag for removal, only to find later that they'd chewed their way out. The special sauce that allows this chemical breakdown to happen lives in their gut bacteria: a symbiotic ecosystem that is an amalgamation of chemicals, enzymes, and bacteria, just like humans' gut biomes. It's unclear which specific elements of the caterpillar's gut are the driving force behind the digestion, but researchers found that the breakdown slowed when certain bacteria from the gut were isolated from the caterpillar itself, indicating there is a more intricate process happening that involves host and gut biome interaction. Breakdown was the same when the larvae were simply smashed and smeared on plastic.

Similarly, some bacteria, algae, and fungi have proven themselves to have an unexpected appetite for plastics. As plastic waste grows, organisms are both suffering and adapting; bacteria and microorganisms are highly adaptable and evolve much more quickly than their larger counterparts. Many natural food sources are becoming scarce because of pollution and environmental degradation directly tied to our garbage-producing world, but in some environments, plastics seem to be a potential additional food source that bacteria are learning to consume.

Discoveries are being made and the field of bioengineering is expanding in an effort to find naturally occurring, cheap, or free abundant methods to dispose of plastic waste. From Japanese scientists discovering a naturally occurring species of PET-eating bacteria to a lab formulating a super-enzyme that degrades plastics on turbo speed, people are working hard to develop and scale up ways to deal with growing amounts of plastic waste. Inventors and scientists imagine creating plastic-eating bacteria that could be sprayed over landfills, contained facilities, or even into the ocean.

An aspirational comparison that regularly comes up in the discussion of plastic-eating organisms is that of certain bacteria's role in oil spills. It's been documented that a group of naturally occurring oceanic bacteria can consume crude oil. This is often given as a shining example of nature doing the dirty work of human error. But when it comes to the same scenario occurring with plastic, there are some big differences.

Unlike plastics that are engineered from petroleum, crude oil is an unprocessed substance stored in underground reservoirs that naturally seeps up from the ocean floor. While humans are responsible for a large portion of oil entering the ocean through spills, drilling, and accidents, there is a vast amount of oil that is always entering into the water by way of these naturally occurring leaks. The oils we use, such as natural gas and diesel, are derived from different levels of light or heavy fuels within crude oil; each of these contain different amounts of chemical compounds, some of which are easier to biodegrade. Unlike sudden disastrous oil spill events, these naturally seeping oils aren't concentrated in a huge surface spill, and little yellow ducklings aren't being washed in Dawn dish soap, so we don't have much awareness of the more natural ecosystem of crude oil.

Because fossil fuels have been around for upward of three hundred million years, their microbial counterparts have been using them as a source of nutrients for just as long. Different microbes consume different parts of the oil, and the heavier the oil is, the fewer available nutrients are accessible to be broken down. The mere presence of oil-eating bacteria is not a recipe for magical cleanup. Many people have suggested that adding such bacteria to ocean oil spills will enhance the cleanup effort: the more the better. However, there is no evidence to show that introducing additional oil-consuming organisms helps during large spills, and it's a logistically unfeasible approach to managing such disasters. Factors including temperature, oxygen levels, currents, quality of oil, and types of microbes all influence how efficiently oil is degraded, and often with wildly different results. For example, "natural processes, including biodegradation, removed 99.4% of the crude oil spilled in Alaska's Prince William Sound in 10 years (NOAA), whereas oil spilled in the Kuwait desert is expected to remain for centuries."[26]

But Perhaps That's Not the Answer We're Looking For

Marine oil-consumers digest the oil and break it down into carbon dioxide and water, using the rest for energy. That's all well and good, but we don't exactly know what potential toxins will come out the other end of plastic-eating bacteria. While the idea of an eco-friendly option to undo our bad habit is idyllic, the reality isn't simple when it comes to erasing our plastic footprint. Much like in any bioengineering, or when introducing a species to an environment, there are risks. In theory, these organisms could be used in large-scale industrial capacities, landfills, recycling settings, or in the ocean; however, taking any organism out of its natural habitat and introducing it large-scale into a new environment is a potential recipe for disaster.

In 1937, Australia made a regrettable decision. Inspired by the success of Puerto Rico in its employment of the cane toad, *Rhinella marina*, to devour the cane beetles decimating their sugarcane crops, Australia thought it wouldn't be such a bad idea to get some toads of their own. Bred from just 101 toads (number 102 died on the ride over from Hawaii), 62,000 little toad babies were released into the wilds of Australia, off to do their government-decreed duty. Try as they might, the toads were not good at their job.

Clocking in at a hefty three pounds on average, these guys aren't the best climbers the toad world has to offer, and with their assigned meals living at the top of sugarcane plants, the toads found themselves on a different diet: anything they could fit into their mouths, including but not limited to mammals, reptiles, bugs, cigarette butts, each other, feces, and spiders. With females breeding twice a year and producing up to thirty-five thousand eggs in one breeding cycle, the country was no match for their new pet. Now at a population of around two hundred million toads, Australia's biodiversity has been irreparably damaged because of the toad's introduction. They compete with native species, they release a poisonous toxin, they carry diseases, and they consume beneficial species.

The New South Wales government has a web page dedicated to instructing its residents how to kill cane toads. Acceptable methods include "stunning followed by decapitation" and "gassing with carbon dioxide for >4 hours"; however, citizens are banned from "rapid freezing or cooling followed by freezing."[27] While people will most likely not be collecting and decapitating PET-consuming fungi, the introduction of an organism to a new environment with the explicit intention that it do its assigned job is not something to decide lightly.

There is incredible and fascinating potential in this new research and discovery. On a basic level, it shows how amazing evolution is. For all intents and purposes, plastic is a foreign substance that shouldn't be in anyone's diet or habitat, and yet now we have organisms that will actually eat the stuff. I have such appreciation for the bacteria, insects, and fungi of the world that have so often adapted to the harshest niches on the planet throughout its history—it's truly incredible and a hopeful sign against the background of current levels of species extinction. If you poke around a little on the topic of these plastic-eating organisms, you'll see a lot of head-

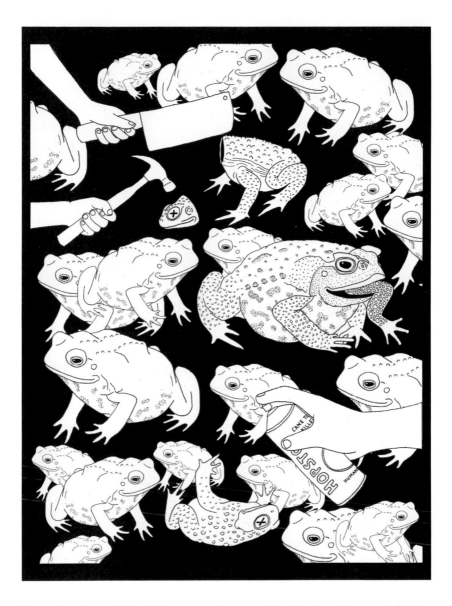

lines about bacteria "solving our garbage problem" or "mushrooms fighting pollution," and there is some level of potential truth to this. But we have a tendency to highlight these shiny, quick-fix answers to our epic waste problem: plastic-eating fungi, reusable shopping bags . . . recycling.

Developing engineered enzymes inspired by nature or breeding one trillion waxworms to consume all the nonrecycled Coca-Cola bottles could be a step toward addressing some of the waste we're generating. Regardless of the approach, many perspectives and solutions will need to be combined to make a dent. But humans have a great mind trick of granting ourselves permission to continue a bad habit if we can rationalize a simultaneous solution that appears to neutralize it—even if it actually doesn't. If a bacteria can eat a bottle, it's not a problem to use that bottle.

These miracle solutions can't let us lose sight of the actual solution: reduction and prevention. Plastic accumulation is not slowing down—it's actually quite the opposite, with plastic waste projected to double by 2050. I'd love to toss a bottle into a pit of larvae and watch them eat it all up, and I can only hope that day comes for the sake of it being cool to see. But there will never be enough wax moths or bacteria to play catch-up with what we've already generated, let alone what we continue to toss on the pile on a daily basis. Bioplastics, compostable cutlery, fungi: it's all useful. But it's also a way to hold on to a lifestyle we desperately want to continue granting ourselves the permission to have.

I don't drink soda, not out of any feelings of elitism, but from a hatred of carbonation, so it's easy for me to say, "Let's just stop manufacturing soda." But that's not going to happen. The people have spoken, and they love Coke, but we all need to acknowledge that sometimes things need to shift in a way that's unlikable to some of us. Corporate greed is the winning party in the game of plastic pollution, and until that changes (possibly through extended producer responsibility), oil companies will benefit from plastic production, plastic production will benefit from consumer responsibility of recycling, and we will ultimately all suffer from it. I don't have a magic bullet solution up my sleeve, but accountability for corporations is the biggest way things will change. It doesn't hurt to change things in your lifestyle to reduce plastic, whether that's buying more bulk goods, starting to use a reusable water bottle, or generally being more aware of what's coming into and out of your ownership. Your choices matter, even if it looks like they don't make much of a dent.

Chapter 9

PAPER

When was the last time you bought a newspaper or subscribed to a daily or weekly delivery of one? I have pleasant memories of receiving our daily local paper, the *News & Observer*, in the late '90s and early 2000s: reading the comics every morning—particularly the color ones on Sundays—and quizzing my parents on the world's weather for the day. While walking around my parents' neighborhood recently, I saw the Pulitzer Prize–winning newspaper sitting on the curb waiting to be brought in, and it was shockingly slim—*maybe* twenty-five pages total, rather than six or seven fully separate sections. Since the early 2000s, the staff at the *N&O* has dropped by 75 percent, and the paper has only 50,047 print subscribers out of a metro area population of roughly two million—a 65 percent drop in subscribers since 2014.[1]

Last year I enjoyed a glorious four months of getting the *New York Times* Sunday paper delivered because it was one dollar a week, and I wanted to do the crossword puzzle, read the advice columns and the Modern Love story, and maybe cruise through some of the other sections. The other (boring) news and sports I usually tossed, burned in my fireplace, or layered into the yard to smother grass after keeping them stacked on the dining table with wishful thinking that I'd get around to reading it during the week.

I felt a lot of guilt getting the newspaper in the first place, unclear on how newspaper printing worked . . . was one extra paper printed just for me? Was one less put into a coin box on the sidewalk—and do they still have those? Did someone stop subscribing the day I started mine? I was tossing so much paper that I clearly was not making the best use of, but I also really love having a physical object to read from rather than spending more time consum-

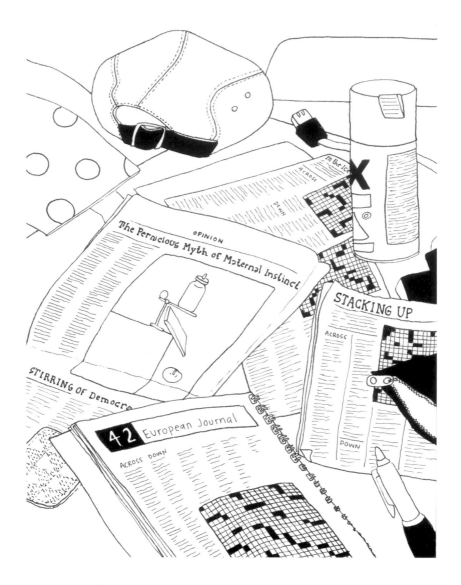

ing information on my phone or computer. It was one of those moments of wondering if my consumer choice of ordering one newspaper per week for four months actually meant anything. I figured, no. I was happy to get to do the crossword every week, and the sum of my paper footprint is far greater due to the books I've written and not due to my Sunday paper. I'm already in deep with feeling responsible for cutting down trees.

My 2022 newspaper subscription was not in line with the US's major decline in printed news, which went from a circulation of 55.8 million daily US newspapers in 2000 to just 24.2 million in 2020.[2] Almost a third of newspapers have gone out of business in the last two decades (a total of 2,500 as of June 2022), with an average of two closing per week.[3] If they haven't closed, local papers have been stripping down to bare bones to save money due to reduced advertising income and higher cost of printing—creating "news deserts," where local news is underreported, accountability decreases, and often stories are syndicated from larger news outlets.

This is good news for reducing paper waste and bad news for . . . news. But despite the move to digital, paper and cardboard still account for 23 percent of municipal solid waste in the United States—nearly as much as plastics, metals, and glass combined.[4] Paper production is the second-largest wood-consuming industry after building materials, and the manufacturing process is a rough one. Turns out that to transform something natural into something not much different requires a lot of steps. Like all waste generation, each stage of the papermaking process, from sourcing to production to disposal, is complicated, with a million variables going into each one. Paper made at one facility may have a wildly different footprint than another in the same town or state or country.

Corrugated boxes make up the largest percentage of recycled material, followed by mixed paper products and newspapers. Annually, an estimated one billion trees-worth of paper is tossed away in the US; 2.65 billion Christmas cards are bought, 141 rolls of toilet paper are used per American (this is crazy), the average US office worker uses ten thousand sheets of paper, and eighty-five billion tons of paper waste is generated globally. We seem to love writing on paper as well, with eight million trees used to make pencils, and nine billion pens tossed yearly. Corrugated cardboard boxes are the most ubiquitous shipping material—drive down your street and you'll probably see some waiting on people's porches. Depending on who you ask, cardboard recycling rates vary from 69 to 90 percent in the United States.[5]

Today, while thinking through my consumption of paper and plastic, I looked in my kitchen to see what's around. There are cardboard boxes or paper bags for granola bars, crackers, baking soda, almond paste, flour, and sugar. The granola bars from the cardboard box are each wrapped in plastic, and the crackers are in a plastic bag in the cardboard box. Cocoa powder, half-and-half, and butter are in nonrecyclable waxed cardboard. There are multiple cardboard boxes sitting on the floor waiting to be walked fifteen feet to the recycling bin, some from shipping and some from product packaging.

I took a very informal, non-peer-reviewed, one-question survey of whether people felt more guilt consuming and throwing away paper or plastic; 120 of 124 respondents said plastic. I didn't expect that, because, for me, there is an emotional aspect of paper production that doesn't exist in the same way with plastics. This is not to say that one is objectively more or less harmful, but when I really think about it, I have opposite anxieties for paper and plastic. I have sadness on the front end of paper: guilt around the massive amounts of wood, electricity, chemicals, and water needed to create paper and cardboard. But with plastic, my emotional response to the use of fossil fuels or chemicals that will be manufactured in a lab is minimal. I'm not happy that we're turning the remnants of prehistoric animals into plastic bags, but I can see and touch a forest and know what it looks like when they're cut down just so I can order something online. Once tossed, however, the guilt flips. It's easier to assume that that cardboard box will be recycled, while the plastic bag I tossed earlier will eventually choke a sea turtle.

A Single-Use Product

Paper products, unlike plastic, glass, or metal, are rarely intended for long, repeated use. Packaging, mail, toilet paper, paper towels, and grocery bags are all close to single-use. Some of it is less than single-use from the outset: each year more than one hundred million trees are used to print junk mail. I'm always pretty shocked at how much junk mail and catalogs arrive in my mailbox. I can't say the last time I looked through them or even unrolled them before tossing them straight into the recycling. Since the move toward online shopping, it's hard to imagine the catalog-driven mail-order business is booming, or that printed ad mailers bring in much revenue.

Since China banned recycling imports, including paper, from outside countries in 2017, India has become a primary recipient of paper from the US, Europe, and the UK. While paper recycling isn't always a valuable endeavor for the places that export it because it can be energy and space consumptive to recycle with variable profit margins, for India, it's a resource they're lacking and will happily accept. India doesn't have a strong logging industry or forests to spare for wood, so receiving recyclable paper is a much-needed way to access that material. Counter to the decline of newspapers in the US, and even with an increasing number of people with access to internet service and digital devices, India's readership of printed news has gone up in the last decade. In addition to having access to paper for news outlets, India's paper market has expanded with vigor to supply packaging for their increasing manufacturing exports and book printing. In what was formerly the red-light district of Old Delhi, there is now an enormous paper market. The Chawri Bazar is home to hundreds of vendors selling things like books, wedding invitations, and newsprint.

Wait . . . It's Not All Recyclable?

But all that paper will not be kept, and not all paper is created equal when it comes to its recyclability or value beyond its initial use. Drumroll, please, for the list of paper that you thought might be recyclable but isn't!

1) The oily stuff. Paper products that are dirty from food or other contaminates, particularly oils, become nonrecyclable. When an oily pizza box or paper plate with a donut stain is tossed into a load of

DOMINO'S PIZZA BOX FOUND ON THE SIDEWALK
ENCOURAGING RECYCLING A NON RECYCLABLE MATERIAL
AS AN ACT OF "DOING GOOD".
IGNORE THIS GREENWASHING CAMPAIGN!

recycling, the food residue can't be separated from the paper fibers during pulping, contaminating the entire batch of otherwise recyclable material. Sadly, no one wants your pizza grease on their office paper.

2) The short fibers. Napkins, toilet paper, paper towels, and tissues, even when clean, are usually made from recycled paper already. This process shortens and weakens wood fibers with every round of recycling, eventually making them too short to reuse. These guys are very short-fibered already, which is why toilet paper disintegrates so quickly in water.

3) Receipts. CVS is killing the world. We've all seen it—that four-foot-long receipt from drug stores offering every survey, coupon, horoscope reading, obituary, and discount known to man that comes with your purchase of shampoo and a chocolate bar. Receipts are, dishearteningly, almost never recyclable, and they sneakily weasel their way into bins left and right. Usually they're printed on thermal paper—a material that contains hormone-disrupting BPA, a toxic "forever chemical" that doesn't decompose and poses health risks to the environment and people. The amount of paper used to print receipts that will almost certainly end up crumpled at the bottom

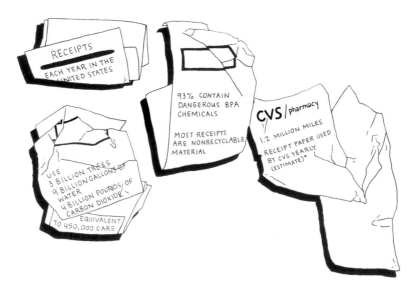

RECEIPTS
EACH YEAR IN THE UNITED STATES

93% CONTAIN DANGEROUS BPA CHEMICALS

MOST RECEIPTS ARE NONRECYCLABLE MATERIAL

CVS/pharmacy

1.2 MILLION MILES RECEIPT PAPER USED BY CVS YEARLY (ESTIMATE)*

USE 3 BILLION TREES
9 BILLION GALLONS OF WATER
4 BILLION POUNDS OF CARBON DIOXIDE
EQUIVALENT TO 450,000 CARS

of your backpack, shopping bag, or the floor of your car adds up to 282,500 tons each year in the US.[6] With more receipts being offered digitally, there is a possibility that receipt use will ease up a little bit. However, just to come clean . . . I am a bit guilty here in not wanting every single store to have my email, so I often still print mine when I have the choice. I look forward to the day when we can opt to have no receipt at all.

4) Your top secret documents. Shredded paper isn't recyclable. This one surprised me! You'd think a smaller version of the same exact thing that's recyclable—a piece of paper—would be equally recyclable since there is no change in the material's physical properties. But like tissues or napkins, when the fibers are shortened, their life spans are shortened, too; or in this case, ended. Recycling sorters' jobs are made (more) difficult when shredded paper is mixed in, introducing a million different useless pieces that need to be sorted out: a game of "try to find the needle in a shredded paper stack." When possible, mixing shredded paper into a home compost bin can add good dry matter.

5) Paper that's accidentally wet or paper products designed to stay dry. I am guilty of putting wet cardboard in my recycling bin. Similar to shredded paper, it seems like it should be OK, since it's the exact same material. But wet cardboard can clog the recycling center's sorting machine, causing major delays or breaking it. The machine gets confused that it's not the weight a light paper item should be and sends it off into the wrong pile. This applies to wet paper in general. So, let it dry out before putting it in the bin—otherwise you might as well just trash it.

The products that are designed to separate wet things from the outside world are, for the most part, relegated to the garbage. Food paper products that *feel* recyclable often aren't, such as paper plates and cups. If they have that slight shine to them, it's because they're coated in plastic. It feels almost pointless sometimes to get paper cups that have the illusion of being more sustainable when really their destination is the landfill, like their plastic versions. Wax-coated cartons of orange juice or broth boxes with foil lining are not accepted in most municipal pickup services, as they require more energy to recycle than other types that are a single material.

Not All Recycled Paper Is "Good"

There are ways that the industry is trying to make progress toward sustainability on both the production and disposal end of things. Sustainably managed forests can serve as carbon sinks, and when

maintained properly they can continue to provide timber while not fully degrading the environment there or in its surroundings. So, when you're looking for paper products and trying to sort through what's legitimately sustainable versus greenwashed marketing, what's the best option? It's not cut-and-dried, but purely recycled paper is not inherently superior. You can buy recycled paper—there's paper that says "recycled" or "postconsumer waste paper," and sometimes the label will show a percentage of how much is recycled material. But the unfortunate truth is that recycling paper requires a lot of energy and water, much of it powered by fossil fuels, and when done poorly, it can verge on self-defeating.

Because recycled paper loses its recyclability as fibers get shorter with every cycle, harvesting virgin wood is a necessary part of the paper trade. Virgin wood fibers are not always evil. When done responsibly, growing and harvesting wood can be advantageous to the economy of local communities and to the health of the environment. Wood is a renewable resource so long as the land has enough stewardship to maintain healthy growing conditions and the wood is harvested in a way that avoids deforestation.

The leading organization of paper sustainability standards is the Forest Stewardship Council, an international nonprofit that works to monitor and protect paper-harvesting forests. If you've ever seen an "FSC-certified" label on a paper product, it's the stamp of approval that the paper was sourced sustainably, which includes fair worker rights. There are three different types of FSC labels that signify a paper's origins. "FSC 100%" indicates all wood from certified forests, "FSC Recycled" means 100 percent recycled materials, and "FSC Mix" denotes a mixture of sources from certified forests, recycled materials, and/or "FSC-controlled" wood, which is noncertified wood that has a known origin that is verified to be legally harvested.

It's OK to use paper. We don't have to go fully digital. And when you're done with your paper product, it's great to recycle it. When seeking out responsible paper products, look for the FSC label, as well as for products that have not been processed with chlorine—a toxin-emitting operation. I printed out this manuscript (single sided to boot) to edit it, and it made my process so much easier. Next week, I'll toss it in the recycling bin and thank it for its service. And that's OK.

AMAZON AND RETURNS

In 2021, an Amazon warehouse in Dunfermline, Scotland, had a weekly quota of 130,000 items to destroy.[7] Video evidence was recorded by a former employee who explained, "There's no rhyme or reason to what gets destroyed: Dyson fans, Hoovers, the occasional MacBook and iPad; the other day, 20,000 Covid face masks still in their wrappers."[8] For the most part, these items were new and unopened, either having stayed on the warehouse shelves too long or having been returned but in perfect condition. The warehouse space is more valuable to Amazon than the products sitting there, and it's cheaper to dump them into a landfill rather than hold on to them or restock returned items.

Amazon is the largest online retailer in the world, owned by the richest man in the world, and as such, it has some of the largest functional and ethical issues of any company in the world. In a calculation done by *Business Insider*, "Per hour, Jeff Bezos makes $8,961,187—that's roughly 315 times Amazon's $28,466 median *annual* worker pay."[9] The entire system of Amazon is a far cry from ethical, spanning the gamut from exploitative treatment of workers, to data collection for price manipulation, to providing technology to the Israeli military, to taking a strong anti-union position, to product substitution, to surveillance issues with digital voice assistant Alexa. They have an awful track record of employee exploitation, including several worker deaths in their warehouses. The company forces unrealistic expectations of pacing, a denial of basic needs like water and bathroom breaks, intimidation against union formation, and warehouse trailers clocking in at 145 degrees Fahrenheit.

Their waste problem is also, unsurprisingly, the biggest of any retailer. If you've ordered something from Amazon, it's likely that you've received a large box containing a tiny item with maybe a couple of those plastic air pillows tossed in there. In 2021, the nonprofit Oceana conducted a study to measure Amazon's waste production, which amounted to 709 million pounds of plastic alone—an increase of 18 percent from the previous year.[10] The COVID-19 pandemic caused a huge surge in Amazon traffic, as people were confined to their homes and relied on the convenience of deliveries. Between 2020 and 2021, Amazon revenue increased 220 percent.[11]

Prime memberships, which offer fast, free shipping on millions of products, increase the amount of excess waste. Ordering two items with Prime, which are both offered to ship within one to two days, is generally too fast to ship them together in one box, even if it's to the same location, as items might be coming from different warehouses or are packed by different warehouse employees. This requires double the amount of packaging and double the amount of driving to deliver what could have been consolidated into one package. Impatience bred out of instant availability feeds the waste generation problem—if we know we *can* get something tomorrow, why wait until a couple days from now, even if it isn't an urgent order? Because of online tools like Google Shopping, and filters within Amazon and other online retailers, customers can compare items based on shipping cost and speed. Oftentimes a product will be essentially the same price as one that ships for a low fee (slightly more expensive with free shipping versus slightly cheaper with cheap shipping), but our little "free shipping" bell lights up *ding ding ding* when we see it, and we're compelled toward the fast and free. According to a UPS study, 55 percent of American customers who bought from online retailers did so directly because of free shipping.[12]

In addition to the packages arriving at your house quickly and for free, returning them is made just as convenient. Buying ten pairs of pants with the hope that one or two will fit in the sizing gamble is a common practice as malls die and those same stores transition to fully online. As with most garbage, convenience is often a catalyst for waste generation. When your bedroom becomes a changing room that has no limit on the number of items brought in, there is no imposed limit on (imagined) consequence-free consumption. Delivery vehicles must drive the packages back and forth, packaging used to send the item in the first place is tossed, and the items themselves are likely tossed with it.

Amazon's stronghold on the retail market is most certainly bolstered by their return policies, which more often than not are free, no matter the reason for the return. There is effort required to open, sort, repair, and re-stock items that are returned—a process that Amazon isn't likely to have interest in completing, since they're willing for workers to suffer from injuries or death for the sake of speed. The temptations of free shipping and returns breed a monster of waste.

My most recent experience of buying/returning in person was a very different experience than overordering online and returning. I've been wearing the same pairs of pants for many years. I bought five of the same style at the same time and they've served me well, although now two of the five have become shorts by necessity, and they're almost all mended in spots and certainly all stained. I decided it was time to bite the bullet and get new pants, as my patience and pants-finding success rate at thrift stores can be limited. So, I channeled my teen self of twenty years ago and went to the mall with emotional support friends. I revisited the same store where my beloved pants came from and picked out six pairs in the same size to try on. None of them fit, as their sizing has changed. Had I ordered online, I would've potentially sent six pairs to the landfill after trying one on and realizing none would fit. I left with three pairs in the right size and, aside from my support of fast fashion, no fear that I would be tossing new clothes into the garbage. I have since gone back for two more of exactly the same kind.

A couple weeks later I bought some replacement light bulbs that were an unusual size. The ones that arrived were bright white and I felt like my kitchen was suddenly transformed into the worst fluorescent office there could be. I was essentially washing dishes on the set of *Severance*. They weren't sold at the hardware store, which is why I bought them online in the first place, and I wasn't able to return them for a set that weren't as awful, so I mailed them back. Because the package was opened, I'm sure that sending them back sent them straight into the landfill. To repackage fifteen-dollar light bulbs would take time that companies would rather not spend; it's cheaper to just throw them away. Even though they functioned, I hated them. I was tempted to keep them in my junk drawer just to assuage my guilt for landfilling, but that made just as little sense if in five years they'd get tossed anyways.

If you're going to order from Amazon, try to order with the option of packaging all your items in the same box to be shipped on the same day, even if it's later than immediate. You probably don't need whatever it is in the next twenty-four hours. Other big box stores are not guilt free and also toss returns, even unopened ones. Sam Hawes from the Hanover County landfill described big-box stores regularly dumping brand-new appliances that had slightly damaged boxes. When buying online, be mindful of what you buy with the intention of keeping it, not the intention of trying it out. If there is an option for you to go in person or order clothing from a site that has a brick-and-mortar location where you can return something you got online, that's a better option that will more likely keep it out of the garbage.

Chapter 10

TEXTILES

One of the most dramatic and tangible shifts toward the culture of disposability is that of fabric and fashion. The amount of yearly textile waste has exploded from 1.76 million tons in 1960 to 17 million tons in 2018.[1] The fashion industry is responsible for 4 to 10 percent of global carbon emissions, second only to oil.[2]

Within two generations, the prevalence of sewing as an everyday skill that people, primarily women, were versed in has diminished to an all-time low, and the rate of consumption and disposal has reached an astronomical high, with three-fifths of new clothing being tossed within a year of its production.[3] Sewing, mending, weaving, and fiber production are part of almost every culture's clothing history, regardless of fashion or materials available. Making one's clothing was a norm among women in the United States across class and race up until the Industrial Revolution when mass production of textiles became cheaper, and marketing tactics promoting newness and fleeting desirability took hold.

It was common practice to make and remake clothing using what was available. The number of clothing items one owned tended to be fairly limited, even for the wealthy; why buy more clothes when they're infinitely transformable? There was a level of practicality and care taken for the clothing people already had because the value of the objects was obvious to those making them. Sewing was very labor-intensive before the advent of the sewing machine, and almost nothing was wasted, so when a garment was torn or worn through, it was repurposed into a new garment, reinvented to meet the current fashion trend, or repaired using mending techniques and scrap cloth. Adult pants with rips beyond repair were turned into children's pants, furniture was reupholstered with remnants, and ripped double sheets were hemmed down to fit single mattresses.[4]

The Beginnings of Fast Fashion

At the turn of the century, a perfect storm of production and advertising brought the first wave of what would later become fast fashion, turning away from homemade high-quality clothing to favor cheaply made seasonal trends sewn by underpaid immigrants. Manufacturers began producing synthetic fabrics that were cheaper to produce and buy. Rayon, invented in the mid-1880s, was the first artificial fiber. When we think of artificial fibers, plastic is often what comes to mind, but these early textiles were derived from wood or plant cellulose—a process that involved little waste but required substantial time. The "father of the rayon industry," French chemist Hilaire Bernigaud, count de Chardonnet, worked to first commercially produce the material outside the laboratory in 1891.

Beyond Chardonnet, rayon innovation made its way through several chemists and companies, eventually popularized and scaled to mass markets by Courtauld & Co., a British silk company. Now under the name "viscose rayon," methods of production shifted from a single input material (cellulose) to a very chemically intensive process, involving caustic soda, carbon disulfide, sodium hydroxide, and a mix of salts and acids, resulting in a silk-like fiber or cotton replacement. While those chemical names most likely mean nothing to most of us, it's abundantly clear that the material is several times removed from its natural state, and there are significant environmental impacts of those chemicals released during the process and as by-products.[5]

Because rayon's base material is naturally derived cellulose, it falls into the category of *artificial* rather than *synthetic*, like its successor, nylon. Nylon, a silk mimic, was a turning point for fashion as well as waste in the fashion world. DuPont, a chemical company steeped in a sinister history, invented nylon in the 1930s and debuted the material in stockings (now colloquially known as "nylons"). In a calculated advertising move, this marketing made the word *nylon* synonymous with a daily household essential that, post–Great Depression, was a symbol of progress, promise, and economic growth, as well as an ability to move away from imported Japanese silk during World War II.[6]

Change in quality came at every stage of the process, creating a more affordable but less durable product. The rise of mechanization, in tandem with the increase in rather savvy advertising, led consumers to associate newness with social status. Printed advertising geared primarily toward women had greater distribution, showing a shiny new life awaiting them for the price of a new oven or dress (or a pair of nylons), all of which would break or go out of fashion soon enough. Disposability became an outward-facing measure of success; creating waste meant having money. It was a newly branded mentality—now a greater number of people could attain what was once only available to the very wealthy: novelty, trendiness, and excess; that was achievable thanks to the rapid-fire, cheap production of goods to be thrown away.[7]

Planned obsolescence isn't limited to the technology sector—it's the ruling force behind fast fashion. Fast fashion heavily relies on synthetic materials to generate cheap products in mass quantities, and in the process it generates a mind-blowing amount of waste. It goes back to the question raised about cell phone designs: Was the cheaper material one that was intended to be superior, achieving certain properties that natural fibers couldn't, with the unintended result of having a shorter life span, or was the product intentionally designed to break down quickly?

In almost all cradle-to-grave models of production, there are problems with manufacturing and disposal that apply to all kinds of fabric, natural and synthetic. Textiles aren't, for the most part, sustainable on either front, and they're one of the worst offenders of excessive waste. Manufacturing methods are energy and resource intensive, production is highly polluting, and recycling is nearly impossible to do on a large scale that could keep up with the rate of disposal.

Comparing the wastefulness of the two categories of fabrics, natural versus synthetic, is a very difficult task. Like most materials that have varied resource options, there are advantages and disadvantages to both.

Synthetics: The Wonder Fabric

For decades, nylons dominated the fashion market and coincided with the development of more synthetic materials. Some textiles had additives (including metals) to make fabrics more appealing but less durable, which increased profits by increasing both purchasing and turnover. Textiles like nylon or polyester are derived primarily from fossil fuels, using about 1 percent of the world's extracted crude oil, contributing to problems with extraction and refinement, and perpetuating reliance on nonrenewable resources.[8] Production requires chemical treatments and dyes to prepare them for use. Global fast fashion giant Shein was found to have products with twenty times the amount of lead that is safe for children, as well as phthalates, which are endocrine disruptors and "forever chemicals."[9] Synthetic fabrics allow fast fashion to produce at lightning speed; every year Zara puts out two clothing lines a month![10]

Not all synthetics are bad, and in fact, they have created ways to provide practical improvements to modern life beyond simply rapid production.

There are reasons beyond cost that influence what materials go into some clothing. Synthetic fabrics perform in many ways that natural fabrics cannot, serving important purposes for casual and technical settings. Synthetics can be more waterproof than their natural counterparts, stretchier, moisture wicking, and more stain resistant—qualities that provide protection in medical settings and high performance in outdoor gear. I went on a multiweek backpacking trip once that was pretty miserable because it rained every single day. I was ill prepared and had packed mostly cotton clothing—a big no-no in the outdoor world, where the safety adage is "cotton kills," since cotton stays wet once wet. I was given some oversize synthetics to tide me over.

We rely on synthetics all around us, and similar to the ways we grow and distribute food, textiles can't be judged solely by their materials. Natural isn't always necessarily the lesser of two evils, and has merits and problems of its own.

Natural Textiles

Unlike many daily, disposable objects that cannot be manufactured with solely natural materials (your cell phone, appliances, or medical equipment), fabric can be fully made with natural materials like cotton, wool, silk, linen, or jute. And let's not forget the animals—wool from sheep, alpacas, and goats, and silk from silkworms. But just because these materials grow from the ground or have renewable sources, natural textiles are by no means flawless; and like all livestock or crops raised on a large scale, they can be grown unsustainably and become hugely wasteful.

The benefits to using natural textiles are many, including the potential for sustainability when done right (or as "right" as something can be done on a large scale). Animal fibers, such as wool, can be produced in ways that promote sustainable grazing and a renewable resource through yearly shearing. Wool is very durable, warm, water-resistant, and absorbent (and if you ask me, itchy), while silk is more delicate and soft. Cotton, who we all know and love, is the fabric you might think of when you think of fabric. It's in our daily lives maybe more than any other natural fiber, in our T-shirts, bandages, and jeans. When burned, chemicals from natural textiles (primarily just dyes that could be chemically derived) are released on a much smaller scale than their synthetic counterparts.

The downsides of producing natural fibers exist on a large scale, and it's hard to make a direct comparison to the downsides of synthetic fibers because the problems are quite different in nature. The animals used to produce natural fibers can be irresponsibly grazed in ways that degrade soil quality, or they can occupy excessive land and resources. Plant-based textiles, like cotton, use an immense amount of resources: land, water, and pesticides. Manufacturing one pair of jeans uses an estimated two thousand gallons of water, for example.[11] Originally cultivated in the United States by enslaved people, cotton is now a huge, mechanized production that often uses genetically modified species. Cotton is often sent overseas to be spun and sewn in factories with low-wage workers where fast fashion brands seek out the cheapest labor they can find, outsourcing to countries like Bangladesh or Vietnam.[12] So, while the negatives are different, they're no less negative for it.

Recyclability

Textiles' disposal must be taken into account just as much as their production when weighing the pros and cons of both. All textiles degrade over time, but the by-product of synthetic degradation is arguably much worse, shedding millions of microplastics—those tiny particles polluting waterways and being ingested by humans and animals alike. There are problem-solving efforts in place to reduce the ever-increasing issue of textile waste, but they are far from perfect. Polyester, a fiber that is likely woven into a large percentage of the clothes in your closet, is often made from recycled plastic bottles. In theory, this is a wonderful idea—using plastic to produce something wearable and functional—but in practice, it diverts the plastics out of a potentially closed-loop system into one that stops when the garment is no longer worn. The plastic fibers can't be reclaimed, and with every wash, microplastics enter the water system, with some fabrics shedding seven hundred thousand microfibers per wash cycle.[13]

There's a push to increase the technology and efficiency of textile recycling, but it's one of the more challenging materials to recycle. Most clothes are made of mixed materials—natural and synthetic blends—with zippers, buttons, or other nonfabric elements attached, as well as inconsistent chemical dyes. Generally, in recycling practices, like goes with like: paper with paper, plastic with plastic, and so on; but when the source material is unidentifiable and variable, its recyclability declines. The basic principle of recycling is to transform the original materials into something identical or quite similar, thereby providing source material for production. In this case, cotton fibers and synthetic fibers must be treated differently to be spun into usable fibers, creating a practical and economical roadblock to effectively recycle the materials. The economic value of recycled textiles is not worth the time and money it requires to recycle them.[14]

Is Thrifting the Answer?

A much-touted solution to reducing personal clothing waste is patronizing thrift stores, but there is an ironic history to their place in consumerism. Goodwill and the Salvation Army established thrift stores in the 1890s to make money to support their Christian ministry programs. It was commonplace for Jewish immigrants to sell worn clothing on the streets, but because of strong anti-Semitism, their efforts were often met with animosity. People were much more apt to buy clothing once it was seen as being under the legitimacy of "do-gooders." This change brought with it a mental shift toward virtuosity in giving to thrift stores; one could acquire more clothing or household goods with the assurance the old ones were being put toward a worthy and ethical cause.[15] What we now see as one of the solutions to reducing our consumption was once a seemingly justified reason to increase consumerism. The altruistic and eco-friendly nature of thrift stores serves as a double-edged sword of guilt-free consumption.

Thrift stores today are an excellent choice to combat patronizing fast fashion, reducing personal waste by not acquiring new clothing that will be tossed in six months. But what actually happens to those clothes when you altruistically drop them off at the thrift store donation center? They're usually thrown away.

Thrift stores receive far too many items to fit in their stores, and only a small percentage of the donations received are in good enough shape to put on the sales floor. According to Maxine Bédat, author of *Unraveled: The Life and Death of a Garment*, 80 percent of donations are unsold.[16] Where they go varies between organizations and locations, but much of the time the clothes are shipped overseas or tossed into nearby landfills. Known in

Kenya as the "clothes of dead white people," exported clothing is plaguing countries such as Ghana, Kenya, Chile, and India, among others.

What Really Happens to Our Discarded Clothes

Chile's Atacama Desert has become a mass graveyard of textiles from counties all over the world. The port in Iquique, Chile, is the hub of textile imports, serving as a convenient spot for countries to dump their sixty thousand tons of unwanted clothes because of its tax-free imports and exports.[17] Some of the imported clothes are then sent on to other countries, including back to where they came from or to in-country thrift stores or markets; and because landfilling textiles is illegal in Chile, the rest are relegated to the desert. Textile recycling isn't efficient or cheap enough yet to make it worthwhile for countries to invest in when exporting it essentially for free is an option.

Not all clothing is sent to a Chilean desert, though many items that end up there are unworn, with tags still attached. In fact, some unworn clothing is destroyed by its own manufacturers. In 2017, the expensive British fashion brand Burberry was caught incinerating £28.6 million worth of clothing and other goods to maintain its status as luxury. Had those items gone on to the secondhand market or been offered at a reduced rate once the season was up, the brand's elite standing would appear to be devalued, so instead they opted to maintain their fanciness by destroying their own merchandise. Part of the appeal of brands like this is their exclusivity. The brand indicates wealth, status, and power; if that was easier to attain, it would lose its glamour.[18]

Other brands have been caught destroying their own unworn inventory once it's past the season or deemed unacceptable for display: Nike, H&M, Urban Outfitters, Louis Vuitton, Cartier, Victoria's Secret, Michael Kors, Eddie Bauer, Coach, as well as counterfeit sports gear for the Super Bowl and other major sporting events. Bags of Nike shoes with huge slash marks have been found on the streets of New York, H&M is notorious for burning or tossing out massive quantities of clothing that have been slashed or had holes punched out, and New York City has laws requiring the police department to destroy counterfeit goods.[19] Called "witnessed burn," the seized products, usually brand-new, are collected and burned or shredded under the watch of the police. In 2014, the NYPD seized $21.6 million worth of counterfeit NFL gear under the guise of brand trademark protections; a law enacted in 2006 directly correlated with knockoff clothing being sent to aid those devastated by Hurricane Katrina. An attempt at providing relief for those in need resulted in a deeply greedy law that looks out for corporations over people.[20] While women's clothing lines have high rates of production and turnover, leaving a lot of excess, menswear doesn't have the same level of excess, so police seizures of knockoffs create a shortage of available free menswear after natural disasters and in clothing banks.[21]

The practices carried out by private corporations and government-sanctioned laws are not only extraordinarily harmful to the environment on the disposal end, but are also a direct cradle-to-grave pipeline that skips the middle part (using the goods). They are manufacturing garbage. Burberry has claimed that their incineration is well-intentioned and beneficial—generating electricity from the waste. However, the energy created from incineration pales in comparison to the energy used to produce and distribute the items in the first place. Many of these companies have tried to cover their public shaming by making promises to donate goods, have trade-in markets, or by claiming that the instances of destruction were unusual, one-off events. While some might make efforts to donate goods, there needs to be a large shift in equating material goods with wealth to the degree that companies will destroy products before their brands are seen on anyone less wealthy than the elite.

Secret Sourcing

While many brands seeking credit for sustainability use "Made in the USA" as a stamp of goodness, brands often don't know where their materials are sourced from, making it difficult to source responsibly and accurately calculate their carbon footprint.[22] Some of the most greenwashed brands actually have no backing to their claims. But this is a choice. Forming relationships with suppliers is possible with some effort, and it allows companies to choose where their materials are grown or extracted from, in order to adjust their impact accordingly. It might cost more to go with the less damaging supply, and that cost might be passed along to customers, but the more that prices reflect quality, the more other businesses might be inclined to follow suit.

Patagonia is one of the brands leading the effort to adopt sustainable practices by operating a secondhand program, donating 1 percent of its profits to sustainable causes, and working toward carbon neutrality by 2025. The brand also published guides on clothing repair to empower people to maintain their clothing rather than toss it out. Their goals are lofty and have yet to be fleshed out, with unforeseen roadblocks of limited supply and overproduction to keep up with increased demand. One problem: Patagonia is not affordable for most people. It falls into the trap of high up-front cost for a longer life span. Those who can't pay $250 for a winter jacket are tied to cheaply manufactured goods out of necessity. Thrift stores generally don't carry clothing that was designed for a long life span, as the whole point of those items are that people will hold on to them, particularly if brands have lifetime warranties.

The fashion industry needs to learn to adjust their supplier relationships, and at the same time, increase their transparency around their company practices. Fashion brands that churn out new designs fuel our need for constant turnover and produce massive amounts of yearly waste. The fabrication of both synthetic and natural materials has the potential to be improved upon, but without accountability, corporations will not change their existing practices. Frequenting thrift stores by donating and shopping, rather than purchasing clothing from fast fashion brands, is an affordable way to reduce involvement in the linear textile economy while not having to spend forty dollars on an organic cotton T-shirt.

There isn't a solution yet to clothing that is thrown away. The amount of discarded clothing is far too much to be recycled or resold. Cheaply made clothes disintegrate or rip, rendering them unwearable and unrecyclable. There are only so many projects you can do with old fabric before your house fills with quilts, pillows, or rag rugs. With eight billion people on Earth, it's too late to return to the ways of small wardrobes with worldwide mending skills, but adopting the mindset of reuse, repair, and reduction is a step toward breaking the cycle of fast fashion without needing to have wealth.

Part 4

THE OTHER 97 PERCENT OF WASTE

Chapter 11

CONSTRUCTION

I n my former neighborhood, more often than not, when a small brick bungalow was bought, it was immediately demolished for a new, exorbitantly expensive rental to be built in its place. All the new builds going up in the neighborhood and around the city look nearly identical, erasing the history of the 1920s homes that were there only months before. I usually only caught them if I was out on a walk—otherwise they'd go from standing to dumpstered in the course of twenty-four hours. When I saw it happening, I would usually stand and watch the enormous machinery smash the house to bits, as if it was made of LEGOs, with no strategy outside of what I can only assume was: smash as quickly and entirely as possible. There was clearly no intention or plan to save materials from the previous structure, despite the bricks being in good shape, the windows perfectly functional, and the hardwood flooring throughout the homes having historic and monetary value. Plumbing fixtures and lighting were most likely left in place, as were doors, tiling, countertops, and hardware. It, admittedly, looked like it would be kind of fun to smash it if the reality of what was being obliterated wasn't so deeply depressing.

While the construction industry seems like it would be primarily additive—more material being consumed than disposed of—a huge amount of construction is preceded by demolition. The construction and demolition (C&D) sector of waste generation in the United States was six hundred million tons in 2018, twice the amount of household municipal solid waste. C&D debris accounts for 30 percent of global waste production.[1] This year I contributed to the approximately 3.4 million tons of carpet alone that is thrown away each year in the United States, comprising 1.2 percent of the total MSW volume, which is most likely underrepresentative, because much of demolition waste is not accounted for in MSW counts.[2]

It's a sector that has only increased as global development expands. Rapid urbanization without a focus on sustainable city planning or construction techniques that fit within a circular economy has created a pattern of construction and demolition that doesn't prioritize reduction and reuse.

Construction without the Waste

A widely accepted idea of "sustainable development" is that a system adequately fulfills the needs of today's generation without compromising future generations' ability to meet their own needs. This is an umbrella term used to describe economic and social growth as well as environmental protections. People have problems with the term and its political implementations, but when it comes to architecture, the principle of sustainable design is centered on reducing waste, maximizing renewable energy production, and having a greater consideration for the surrounding environment, both urban and rural.

Like with most things related to waste, humans used to be really good at sustainability, but moved away from it with the expansion of industry, and are now trying to find a balance again. Many engineers and architects are working toward developing more sustainable building techniques by focusing on innovative design and using more sustainable materials that are durable, modular, or adjustable to encourage longevity rather than replacement. Duncan Baker-Brown, cofounder of BBM Sustainable Design, wants "cities [to be seen] as material stores for the future," keeping building materials in circulation through the practices of "deconstruction" and "reconstruction."[3]

The concept of deconstruction, also known as "construction in reverse," has been around for millennia, dating back to ancient stone buildings whose stones were reused over and over. I love shopping at reuse centers, like Habitat for Humanity's ReStores, and have outfitted my house with cabinets, paint, doors, light fixtures, and a sink vanity from my finds there. This is a sector known as nonstructural deconstruction: using materials that are not integral to the structural integrity of a building. This is the easiest, most accessible form of deconstruction—useful for home projects, handy people, and small-scale builders or contractors.

Structural deconstruction to reclaim materials such as brick, drywall, framing, lumber, and shingles is harder to implement on a large scale, as there are a couple of big issues that make it challenging to put this into practice. Communication platforms and physical spaces are needed to facilitate the material exchanges dedicated to helping builders and demolition services keep track of what materials are available and in circulation. For one thing, you need to be able to store the materials you've deconstructed before they're picked up for reuse, which means you need land and facilities. Insurance companies and builders alike can be wary to give the OK to materials that have unknown levels of quality, either by virtue of being used or old enough that their product details are unavailable to research. There are more unknowns and risks associated with structural deconstruction. Mold, asbestos, and lead paint are all common in older structures, which often make up a large portion of teardowns. However, a huge proportion of materials are stable and don't degrade in a way that would render them less safe or valuable over time. Hardwood, glass, stone, and brick remain stable and can be reused in their same form, and gypsum board for drywall can be crushed to amend soil. It's estimated that 75 percent of construction waste that is destroyed still holds value.[4]

Given the current lack of infrastructure to make material reclamation practical, the acquisition of used building materials requires intention, research, and commitment to sustainable building. For moneymaking (often gentrifying) endeavors, such as the rentals in my neighborhood, those practices aren't accessible enough to beat out the appeal of speedy construction, even if it produces an inferior space.

The advantages to using reclaimed materials and leaning more heavily into deconstruction rather than demolition are numerous and enormously beneficial. From a sentimental and aesthetic perspective, precious resources, such as beautiful hardwood floors or handmade porcelain tiles, could have a longer life and imbue spaces with character and history. From an environmental perspective, not only does reuse keep materials out of landfills or incinerators, but it also prevents excess products from being manufactured using energy-consumptive virgin materials.

MAKING NEW MATERIALS

The effects of some building materials are obvious—we can all see the potential ramifications of cutting down trees to produce lumber for framing—but there are stealthy sources of waste, too.

Cement and concrete production alone generates 8 percent of global greenhouse emissions from the high heat required to produce and process it. There is work being done to make cement and concrete a more sustainable industry, but reuse is by far the most effective and efficient way to create less waste. Injecting the CO_2 produced during the making of the product directly back into the cement itself keeps it out of the environment and acts as a carbon sink—a place where carbon can be indefinitely stored rather than released into the atmosphere.

Reused brick wall on the Resource Rows

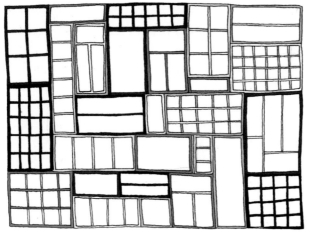

Reclaimed
windows on the
EU headquarters in
Brussels, Belgium

International Approaches

On a very public scale, the European Union headquarters, the Europa, in Brussels, Belgium, has made reclamation the face of its building, quite literally, with a façade made of all reclaimed oak windows. For those who aren't designing internationally important buildings, several European companies are working toward making reclamation more attainable by establishing searchable databases that list materials that are available for use, and one Copenhagen-based architecture firm, Lendager, experimented with designing and building a housing complex made from upcycled materials. Resource Rows features a façade of mixed one-by-one-meter sections of reclaimed brick wall held together by their original mortar and put together like a patchwork quilt. The bricks were repurposed from old buildings within the city, as were the lumber and windows used in the complex's community garden. By repurposing just 10 percent of the total materials, the building saved 29 percent of the total CO_2 emissions that new materials would produce from the same project.[5]

In Tokyo, an experiment in subtle deconstruction took the shape of dismantling a forty-floor hotel, the Grand Prince Hotel Akasaka, from the top down with no explosions, wrecking balls, or publicly observable demolition. Using jacks and regenerative cranes that generate electricity while in operation to power other machinery in use, the building was internally dismantled one floor at a time. Once one floor was emptied, temporary support columns were removed and the floor above was jacked down, all while the roof of the building remained intact, producing the visual effect that the building was mysteriously shrinking. Almost all the materials were salvaged, and the technique greatly reduced noise pollution as well as debris and dust in public spaces.

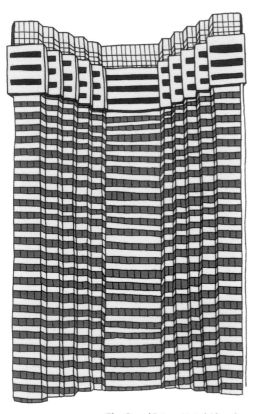

The Grand Prince Hotel Akasaka

Printing Houses

While demolition strategies are becoming more creative to meet the goal of less waste, building innovations are harnessing new technology and old principles to cut down or eliminate resources like cement and timber, opting instead to use novel materials. Additive manufacturing is the base of 3D printing, allowing unique, computer-generated designs to be fabricated into 3D objects by a machine. The world of 3D printing has been upping its scale to manufacture entire homes in an effort to increase sustainability, reduce environmental impacts, and create more affordable housing options. More than one hundred prototype houses are already being inhabited by formerly houseless people in Austin, Texas, as part of an initiative to provide low-cost housing.

Having a mechanism to produce on-demand materials would eliminate a huge amount of transportation cost, energy, and labor by enabling exact shapes and sizes needed to fit custom designs that ordinarily produce waste on site. All those extra wood scraps and shingles that you didn't use from your bulk purchase . . . they add up in the dumpster. Because 3D printing can use a myriad of materials and easily lend itself to quickly shifting designs, the possibilities of efficiently integrating design elements are vast. Growing research into using recycled plastics as 3D printing filament has the potential to reduce waste further.

There are many appealing and compelling aspects of 3D printing as a tool for large-scale construction, yet there are also people not so certain of its sales pitch, including me. I don't love the idea of skilled labor being replaced by machinery, but I think the integration of these ideas could lead to some very interesting, unique, and functional spaces. The application of this technology is still very early in development with an uncertain future and its fair share of critics, but even if it has no viable future as a feasible method of affordable, sustainable production, it's a sign at the very least that people are thinking of ways to shift mindsets and practices toward a better way to build.

Back to Basics

In contrast to fancy robot-printed homes, older principles of design that take advantage of Earth's natural heating and cooling are also resurfacing in modern architectural practices. These buildings are designed to prioritize passive systems that require far less energy consumption than designs that do not take those natural elements into account. Sunlight helps heat a home and its water tank while providing natural light; wind can generate energy; steady ground-source temperatures can efficiently provide heat to a home; and intentional orientation can increase the energy efficiency of a home drastically.

A couple years ago, I took a natural building class where we learned to build with cob (a mixture of clay, sand, straw, and water), and construct straw bale houses and living roofs. Almost none of the materials were purchased, aside from plastic liners for the roof and straw, which is a cheap renewable resource that serves as both the insulation and structure of the walls. We used salvaged windows, clay from the property, and sand from nearby, and we mixed the cob by stomping on it. Alternatively, a small electric cement mixer will do the job, since some people, like me, do not love the stomping part.

An extreme version of this low-waste, high-efficiency lifestyle is an Earthship home. First developed in the 1970s by Michael Reynolds, Earthships are designed to incorporate existing waste and generate none of their own. Taking inspiration from the natural materials around him, as well as responding to the energy crisis of the 1970s, Reynolds built some of the first Earthships in Taos, New Mexico, using adobe, mud, old tires, aluminum cans, glass bottles, and reclaimed windows. Used tires make up the base of most Earthship walls, a resource that is more than plentiful and a big source of pollution; they can be used without any modification or further processes and are small enough to be moved by hand with some elbow grease. Heavy and load bearing, tires are filled and surrounded by rammed earth, making them virtually indestructible. They also serve as a heat battery, warming up during the day, storing the heat in the wall mass, and releasing it slowly during the night.

Building does not have to be so resource demanding and waste generating when there are so many resources available for reuse. Reclamation is the way buildings were built until the past century, and the method is still used out of necessity in many places, driven by limited access to resources, monetary or physical. Around the world, people are turning to reclamation as an opportunity for design, environmentalism, and economics. Reuse is an opportunity for creativity—being limited in what materials one has forces you to think in ways that you might not if every resource were available to you. That is not to say that innovation is only done when resources are limited or that limited resources is inherently a positive thing, but in studying how reclamation has been done for millennia on various scales, we can gain some understanding of how to implement those ways of thinking.

Like with most waste systems, we are in a moment where the start-up to construction sustainability requires a huge influx of effort and money, but in the end produces long-term efficiency and savings. Developing systems of material exchange between builders and manufacturers can create a circular system that doesn't impede building or deconstruction. Affordable homes could be more easily constructed with reused materials, and demolition crews wouldn't need to pay for landfilling. The growing interest in sustainable architecture is promising, but the rapid rise of development is alarming. Rehabilitating or deconstructing dilapidated structures in areas where there are tons of vacant buildings, as a way to build housing where affordable housing is scarce, could be a situation like food waste and gleaning: use demolition waste to fill a need in the housing crisis.

Chapter 12

MINING

Mining in some form or fashion has been going on for tens of thousands of years. Believed to be the oldest mine in the world, the Ngwenya Mine in Eswatini dates back forty-three thousand years. The hematite ore extracted there was used for painting and funerary practices; over time and across the globe, mining expanded to employ different methods of extraction and mine a wide variety of materials. During the Neolithic period, higher levels of salt, gold, flint, amber, and ochre were required to forge tools and pottery, and to sustain the dietary changes of growing non-nomadic societies while amplifying increasing global trade systems. Salt became a defining feature of diets and fetched a high price in turn; in fact, the origin of the word *salary* comes from *salt*, as salt was at times a form of payment to Roman soldiers. Our present-day mining was built upon our ancestors' basic needs, to accommodate our ever-growing appetite for fuel, infrastructure, and technology. Most of the objects around you probably involve mining to some capacity: the aluminum foil your lunch was wrapped in, the metal chair you're sitting in, the lamp and light bulb nearby, your cell phone, and your laptop. Mica in your toothpaste, gypsum in your drywall, talc in your makeup, clay in your brick house, pumice in your cat's litter, copper in your wiring: it's everywhere!

Mining waste is one of the hardest forms of waste for the general public to see; the waste generated to extract the mica flakes for your toothpaste is even more out of sight, out of mind than the landfill its tube is eventually heading to. Unless you're at a quarry or extraction site, you can pretty much only see it in the most extreme cases of mountaintop removal or

giant terracing cut into the side of a hill. The process to extract and process minerals, coal, and metals from the earth leaves behind tons of waste
compared to the amount of desired material that's ultimately used. The
mining industry is the largest emitter of toxins in the United States, accounting for 41 percent of all reported toxins released,[1] and it's estimated
that over fifty million gallons of mining waste pollutes local water sources
daily.[2] There are a couple different forms of waste generated from mining practices: tailings and spoil tips. Tailings are the non–economically
valuable materials resulting from the extraction and separation process.
Spoil tips are piles of waste removed to create the mining site or reach the
desired materials.

The World's Grossest Ponds

Tailings are generally the more troublesome of the two with regards to
toxicity. After chemicals are used to separate the target materials, the tailings produced are usually very fine particles laden with toxic chemicals
that, when mixed with water, are reduced to a sludgy consistency kept in
tailings ponds. *Pond* might evoke the wrong image: a quaint little body of
water with chirping frogs and lily pads; in reality, these sludge piles can be
massive and are almost always inhospitable, gross places. The Syncrude
Tailings Dam in Alberta, Canada, is one of the largest man-made structures on Earth, and certainly the largest dam, at eleven miles long and 289
feet high at its tallest point.[3] For comparison, the Hoover Dam is more than
twice as tall at 726 feet, but much, much narrower at 1,244 feet.

Some waste by-products of mining are bad news; phosphate tailings
contain uranium and thorium, gold tailings are rich in cyanide and arsenic, copper tailings include uranium, thorium, and radium, and taconite
tailings contain asbestos— a smorgasbord of poisons. One form of phosphate mining by-product known as "wash slime" is a sludge that never
dries because of a higher water content, thereby creating a permanently
gross wet slurry with no use.

Retained behind tailings dams, these toxic sludge piles (also known as
mine dumps, slimes, or slickens) have a tendency to leak to disastrous effect because of failing dams. Unlike a traditional dam in a river, which too
can have catastrophic consequences if breached, a broken tailings dam

TRASHY TIDBIT

1962–present:
Coal Seam Fire, Centralia, Pennsylvania

The town of Centralia, Pennsylvania, once had a population of 1,800 and now has less than ten residents and no zip code, thanks to a noxious coal seam fire that has been burning since 1962 with no end in sight. The origin of the fire still isn't clear, but we can now learn from example that one shouldn't build a landfill in an abandoned strip mine and then set it on fire. Sinkholes formed, residents got carbon monoxide poisoning, and smoke billowed up from the cracked streets.

The United States Postal Service took away Centralia's ZIP code, and residents were paid to relocate, save for the few holdouts who are technically illegally squatting in their own homes. Once these residents die, the population will, most likely, remain at zero.

can cause massive damage because of its contents. In January 2019, the Córrego do Feijão iron mine dam in Brumadinho, Brazil, broke and unleashed 3.4 million tons of toxic sludge into the mine's offices, cafeteria, and then into the downstream town, killing a total of 272 people and causing massive physical destruction.[4] Other huge disasters have occurred as a result of dam failures, causing mudslides, flooding of lowlands and rivers, and polluting groundwater.

Unlike tailings, which are generated as a result of material processing, spoil tips are the piles of debris removed in order to access the desired ore. They could include rocks blown off in the process of mountaintop removal, gravel dug up from creating tunnels, manure piled up from ex-

THE BELAZ 75710:
BIGGEST DUMPTRUCK IN THE WORLD

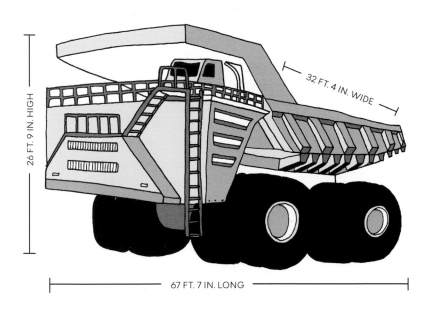

26 FT. 9 IN. HIGH

32 FT. 4 IN. WIDE

67 FT. 7 IN. LONG

1,300 LITERS PER 100 KM
=
0.18 MILES PER GALLON

cavating farmland, or earthen materials dislodged from hydraulic mining. Sometimes the amount of excavated material is enormous, dwarfing its surroundings, at times reaching heights of over five hundred feet, and even earning UNESCO World Heritage Site status at the now-defunct mining site Loos-en-Gohelle, in France.

Politics and Demand

There's the ultimate irony to contend with in the mining sector. Most elements mined to produce EV batteries are mined using fossil fuel–run machinery—with each metric ton of lithium mined, fifteen metric tons of CO_2 are emitted in the process, for instance.[5] In fact, one of the contributing factors to a recent increase in mining is our transition away from fossil fuels to renewable energy: solar, wind, and electric. The shift from fossil fuel–based economies to "green" ones is a steep climb that will require a huge increase in mining—an industry that is paradoxically powered by fossil fuels. Mining activity will have to quadruple to build enough clean energy technologies by 2040 to meet the greenhouse gas emission goals of the 2015 Paris Agreement, an international climate change treaty that's aiming for climate stabilization and prevention of global temperature rise. To reach an even loftier goal of hitting net-zero global emissions by 2050 would require six times more mineral inputs in 2040 than today.[6]

We use the word *renewable* for the technology that will hopefully keep global disasters caused by climate change at bay, and while the energy sources that power those technologies (the sun and wind) will never run out in our lifetimes, the elements required to build them are nonrenewable. The metals and minerals needed to build these machines are finite, and we will run into the same issues with mining them as we eventually will with fossil fuels.

As with coal or oil, there are different purities and qualities of these mined elements. The higher the grade or purity, the easier it is to process and the less energy it requires to get it from the earth and into a usable state. Over time, the higher quality material will deplete, leaving behind the less valuable, more energy-intensive matter. It's a tough trade-off. To get where we need to get in terms of global emissions and climate change, there is a sacrifice. For people who have a glistening view of the shift toward sustainability, the harsh reality is that it comes with a pretty hefty cost to the environment to get there.

There are some problems with meeting the electric demand; the primary materials required to support these technologies are copper, nickel, cobalt, rare earth metals (for magnets), lithium, manganese, and graphite. Like oil or coal, these finite materials are highly concentrated in certain areas and low or nonexistent in others, which will inevitably create some serious global political issues, much like with fossil fuels. The Democratic Republic of the Congo (DRC) and China combined produced 70 percent of the cobalt and 60 percent of rare earth metals on the market in 2019. China has ownership of fifteen huge DRC mines, and a huge upper hand in the industry for nickel and lithium as well, with "about 70–80% of the refined cobalt market and probably half of the battery market."[7]

Mining has the potential to cause civil and governmental unrest in areas with prolific mining industry and wealth, often creating illegal operations and exploitative labor. The DRC's government is tightly connected to the Global North's investment in mining. As the world's largest source of cobalt and home to many more mineral resources, their leadership is in the pocket of corporate interests. The DRC has completely altered its landscape for the sake of extracting cobalt, a primary component of lithium batteries, which is not only challenging to extract, but extremely toxic when touched or breathed.[8] Known as "artisanal mining," a name that might conjure up images of freshly baked bread or fancy cheese, the process couldn't be further from that association. *Artisanal* in this case means individuals, often kids, are barely making a living by dangerously hand-excavating cobalt as well as other substances for the sake of renewable energy. Many of the mine workers have no alternative ways to make money in areas ravaged by mining industry and controlled by militia groups. What kind of sustainability is that? Mining often causes illness to those nearby; endangering human lives for the sake of sustainability is a corrupt way to get there.

Not only is mining endangering the lives of people through dangerous labor practices, but it also increases risks of negative impacts from climate change, including threats to water access. Flooding and extreme precipitation are going to increase with climate change; too much water plus underground tunnels or mountainsides is generally a bad combination that can lead to tailings dam failures, tunnel collapse, and mudslides. Extreme weather already causes 25 percent of dam failures and is likely to continue, even as dams are built in more secure ways.[9] Water scarcity poses a more immediate, persistent threat to mining operations. Lithium and copper require huge amounts of water to produce, and in areas of existing or increasing water scarcity, it will become more challenging to obtain the water needed to process these materials.

Chile produces the most copper and the second most lithium in the world, and is facing a drought that could lead to decreased production from mines. This change in local mining production could increase political and public tension between residential and industrial water consumption as supply decreases. In 2020, between 30 and 50 percent of zinc, iron, and copper was produced in areas undergoing high water stress.[10] I am not an oracle, but I think it's safe to assume that water scarcity is not an issue that will improve in the future, and the best-case scenario would be that it remains stable. In many places where mining is the primary source of revenue, governments impose minimal restrictions, if any, on mining water use, granting companies excessive access to water rights with no requirements that it be reused or properly treated. When facilities are upstream from communities, they pose a health risk to the residents' water quality in the form of limited access to clean water, or adequate amounts of water and increased risk of flooding.

Ways to Clean Up the Mess

Sometimes facing scarcity or impending hardship forces adaptation, and in the case of mining, there is a vested interest by some parties to make mining operations as viable as possible, to maintain local and national economies in areas that are resource rich. Rather than attempt to fight tooth and nail to acquire more and more water rights, implementing tools to reduce evaporation and increase reuse could ease the high water demands of mining facilities. Governments could impose restrictions on companies to adjust allocations that reflect the current amount of local water access; they could require implementing technology to reuse water as much as possible; and they could have strict contractual agreements of where and how tailings dams are constructed and maintained.

Geothermal extraction is a newer form of lithium mining that has very little known environmental impact in comparison to the most common methods of hard rock mining and underground reservoirs. In geothermal extraction, lithium is extracted using heat and movement of water, with no human interventions that rely on fossil fuels.

In theory, the demand for minerals will require a hefty start-up cost that eventually tapers off as economies come to establish renewable energy infrastructure with long life spans. With coal being the primary revenue source for mining industries, there is an economic incentive to continue mining coal; it is a huge money maker. However, as the volume of minerals and metals required to produce electric technology ramps up, there will, unfortunately, continue to be an abundance of mining money to be made—even if or when coal-powered industries go away. Like bioplastics or paper shopping bags, the requirements to sustain the lifestyles we're used to and that more people around the world are adopting make sustainability harder, and substitutions still come with their disadvantages.

So much of mining is not in the interest of producing sustainable technology, and is doing quite the opposite by supporting the fossil fuel industry, but the benefits of an eventual reliance on renewables rather than fossil fuels will hopefully outweigh the damage caused by the intensive mining necessary to build technology on a large enough scale. If building electric vehicles, solar panels, and wind farms is done right, with durability and minimal waste in mind, mining for coal and fossil fuels will decrease over time, and the rate of growth to manufacture those clean energy products will ease.

Enforcing fair-wage, safe labor practices, better water reuse technology, and operations that do not completely destroy the surrounding environment are ways to mitigate what is inevitably just a damaging practice. There is only so much sustainability that can go into extracting a nonrenewable resource, but for companies touting their own greenness, it's a hypocritical move to get there by deeply unethical means that are exacerbating the issue they're trying to fix. It's important to be aware of where your renewable technology is coming from and research the sources of your new technology, whether that's an electric vehicle or solar panels. As of 2021, Volkswagen, Volvo, Renault, and Mercedes-Benz have all been found to source cobalt from the DRC. If you have the financial resources to buy an electric vehicle or install solar panels, you are likely someone who is attempting to improve your environmental footprint. With that privilege also comes the responsibility to not greenwash your own behaviors. Take the time to understand the costs that come with your purchases, and find out which companies are operating with ethical standards for battery manufacturing.

Shifting our waste practices requires thought and intention—and it's not an invitation to throw ourselves into a different kind of waste for the sake of reducing the other kind.

Chapter 13

RADIOACTIVE WASTE

One of the most interesting books I read recently is *The Radium Girls* by Kate Moore, a true story about the women who worked as dial painters for the Radium Dial Company in the 1920s. Spoiler alert: After consistent ingestion of radium paint and frequent contact with the substance on their skin, the radium sickened and killed a lot of them. The ensuing legal battle and community outrage publicly illuminated the hazards and ill effects of radium on human health, and gave me a more tangible understanding of how radioactive substances interact with the body/organisms. Watching the show *Chernobyl* was another extreme and horrifying version of the same realization. Seeing the skin melt off people and reading about jaws crumbling put into perspective why radioactive material is in those big yellow bins with the scary symbol.

The 1986 Chernobyl disaster in the former Soviet Union and the Fukushima disaster in Japan are both examples of the worst-case scenario of nuclear accidents and the power of radioactive material. The Chernobyl disaster released enough radioactive material to create an uninhabitable level of contamination with a radius of thirty kilometers surrounding the plant, known as "the exclusion zone." Upward of 150,000 square kilometers beyond that were contaminated with low-level radioactivity, with particles carried throughout almost all of Europe in the weeks following the disaster.[1] Water in the neighboring Pripyat River was contaminated, affecting drinking water and aquatic wildlife. The trees in the nearby pine forest died soon thereafter, and land animals, both domesticated and wild, suffered from thyroid issues and mutations. Since then, the levels of radioactivity have diminished with decay, and people have returned to their homes in the plant's exclusion zone, which was immediately evacuated after the disaster.

Because of their dangerous effects on the natural world, radioactive substances need special treatment and methods of disposal. There are several kinds of radioactive wastes that require different ways of handling: high-level waste (the classic glowing neon green of *The Simpsons* . . . it's not green or glowing in actuality) and low-level waste (rags or lab rat carcasses that have come in contact with radioactive materials). The majority of radioactive waste is low-level and relatively easy to deal with, but the higher-level hazardous waste requires careful disposal.

Radioactive waste is produced during scientific research, mining, military activity (nuclear weapons), and medicine. Militaries in China, North Korea, France, India, Israel, Pakistan, Russia, the United Kingdom, and the United States have all produced massive amounts of radioactive waste by testing and manufacturing nuclear weapons, with Russia and the United States possessing about 90 percent of them. And outside of weapons, there's also nuclear energy.

A Controversial Energy Source

Those concerned about climate change generally think of renewables as clean, sustainable, and green, and fossil fuels as dirty, finite, and environmentally destructive—so where does that leave nuclear? As of 2022, nuclear power supplies 18 percent of the United States' energy, only slightly less than all renewables combined. As such a major source of power, nuclear falls into an interesting position in comparison to other energy production, with strong advocates for and against its use.

The benefits of nuclear power are many, including very little greenhouse gas emissions and air pollution once in use. With increasing urgency to address climate change, nuclear is gaining more approval as a source of clean energy to move away from reliance on fossil fuels. The public image of nuclear power has been dominated by disasters: atomic bombs, plant catastrophes (caused by natural disaster or human error), and risk of terrorist attack, as well as concern for nuclear storage of spent fuel. While there is fear around its safety, or lack thereof, nuclear power is actually one of the safest methods of energy production, with very few accidents, disease, and deaths directly caused by energy production. Disasters are very rare in the grand scheme of energy production from nuclear, and facility safety standards have only increased over time.

Other sources of energy are far more dangerous, with a recent study finding that a staggeringly high one in five deaths worldwide can be attributed to fossil fuel air pollution.[2] While coal mining has become less dangerous to those working in the mines than it was a hundred years ago, the proliferation of fossil fuel use has created a slew of problems worldwide. Air pollution causes and exacerbates health issues across the globe, while nuclear has no air pollution and almost no occupational hazards directly correlated to the production of power, making it one of the safest forms of energy production possible.[3]

Despite all the positives, nuclear power has been a source of debate since its inception in the 1950s, with the antinuclear movement gaining momentum in the sixties. Opponents have included many environmental organizations and pro–fossil fuel lobbyists—two groups that have very different ideals around energy and climate, but agree on their antinuclear sentiments. Some advocates are primarily concerned with the threat of nuclear weapon manufacturing—a fear strengthened by the Cold War and the attack on Hiroshima. The distinction between harnessing nuclear power to generate energy versus using it to construct nuclear weapons is an important one when considering it as an option among all the other energy sources, such as renewables and fossil fuels. While the fear of atomic weapons is more than valid, a country does not need to have operational nuclear power plants to produce nuclear weapons.

One problem with nuclear power is the process of acquiring the uranium required to run the plants. Uranium is generally obtained from open pit mining, a process than can be very destructive to the environment and must involve chemicals to separate uranium from the ore around it. This mining and processing can leave radioactive waste material behind, and also pose radioactivity risks to the miners.

Despite the pros and cons of this power source, large-scale energy production usually comes down to economics, and in this case it's yet to be seen what's in store for nuclear energy in the future. Nuclear power plants are exceedingly time consumptive and expensive to build, costing upward of $10 billion. In the last several decades, the number of nuclear plants decreased in the United States and much of Europe, which relies on nuclear power as an important source of energy, as plants are reaching the end of their life spans. In the current movement toward green energy, the prospect of investing in nuclear—whether that's making improvements to existing plants, building entirely new plants, or shutting them down in favor of other sources of renewable energy—is yet to be seen.

Nuclear Power's Powerful Waste

While the production of nuclear power plants is very clean and efficient, the plants currently in operation produce two thousand metric tons of radioactive waste every year in the United States alone. Nuclear plants generate 10 percent of global energy, with France, the United States, and China being the world's largest producers and consumers of nuclear energy. In 2016, thirty countries operated 448 nuclear power plants. To make nuclear power, uranium is formed into pellets and put into long metal tubes (the fuel rods), which heat up to power turbines and produce electricity. After four to six years, nuclear reactors use up enough of the uranium fuel within the rods to make them no longer 100 percent efficient; once at that stage, the rods are shut off and removed from the reactor. Spent fuel can often be reprocessed by extracting its uranium and plutonium to reuse in the reactor. Upon removal, the rods themselves are very hot and very radioactive, primarily because of the by-products created from the fission reaction, which diminishes substantially over the first few days and progressively thereafter.

The problem then emerges of how to permanently dispose of spent rods. The current solution is to keep them submerged in pools of water, which both cool the spent fuel as well as create an impenetrable barrier for radioactivity. This is one instance where the Olympic-size swimming pool comparison could actually come in handy: to compare a pool to a pool. Unfortunately, I have no idea how many spent fuel rod pools there are, but I do know that they're generally around forty feet deep, whereas an Olympic pool is less than ten.

The amount of available space in cooling pools is insufficient for the amount of waste being produced, so in the 1980s, dry cask storage was developed as a way to store the waste both on- and off-site. Dry cask storage feels a little bit like a magician trying to come up with all the obstacles they can to uncuff their hands from behind their back—swallow the key, wear a straitjacket, hang upside down, be blindfolded and raised ten feet off the ground. In dry cask storage, the magician trying to get out is nuclear waste. After cooling in the pool for one year, it's surrounded by inert gas inside a steel cylinder, which is bolted or welded shut, surrounded by more concrete or steel. They kind of look like oversize tin cans all lined up in a grid—tin cans that could hold fifty midsize cars.

There is still no permanent solution to nuclear waste storage anywhere in the world (as of 2023). Despite that, nuclear power plants have been operating since the 1970s. Some elements from spent fuel rods have a half-life of approximately twenty-four thousand years (plutonium-239) and two million years (neptunium-237), meaning that after two million years, neptunium-237 will be half as radioactive as in its original state. When conceiving of a permanent spot to store nuclear waste, this overwhelmingly huge timeline must be taken into account: What can be stored for potentially millions of years without damaging the surrounding environment or decaying the infrastructure over time? Homo sapiens (us) evolved from our ancestors three hundred thousand years ago, so it seems like a pretty impossible task to envision building something that's intended to remain stable and unchanged for more than three times as long as we've been on Earth.

Currently, waste is either stored on-site at plants or shipped off-site, but Finland is in the process of building the first permanent nuclear storage facility in the world.

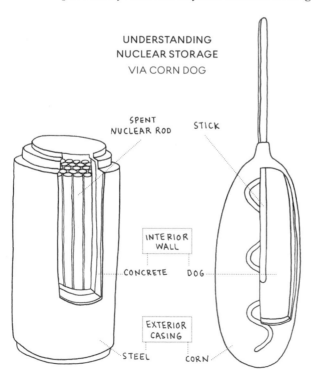

UNDERSTANDING
NUCLEAR STORAGE
VIA CORN DOG

SPENT
NUCLEAR ROD   STICK

INTERIOR
WALL

CONCRETE   DOG

EXTERIOR
CASING

STEEL   CORN

The fittingly named Onkalo, which means "pit" in Finnish, will be an elaborate storage system involving copper casks, robots, tunnels, and fans. They're serious about the longevity of its stability. The casks "must remain undisturbed for 100,000 years, even as the warming climate of coming centuries gives way to the next ice age."[4] I'm not convinced humans will still be on Earth by the year 102,203, so it'll hopefully outlive us all.

The US government attempted to secure Yucca Mountain in Nevada as a permanent storage site for the country's radioactive waste, but after $19 billion was spent to research and develop a test tunnel, it came to a halt during the Obama administration due to geological issues. It turns out that the site is in a potentially seismic area near a large aquifer, and the mountain itself is made of porous, cracked rocks—not a good material to contain radioactive substances, and information you'd think they would have gathered before tunneling. Political backlash from both the government and Nevada residents was actually the main roadblock to completing the site, with the governor of Nevada filing a legal motion in 2022 to permanently stop the licensing of the Yucca Mountain repository. Nevada has no nuclear facilities and would be subjected to trucks regularly bringing in nuclear waste through the state—a potentially dangerous endeavor. Many feel that it would make their state a dangerous experimental dumping ground. In choosing a site, officials also failed to take into consideration the significance of the site to the Paiute-Shoshone Tribes, who have held Yucca Mountain as a spiritually and culturally significant site for thousands of years.

It's yet to be determined what solution will develop for the future of nuclear waste. The methods we currently have are certainly unsustainable, and the ever-existing argument about the ethics of nuclear power plants as an energy source persists. This element of waste is absolutely out of your hands as a consumer choice. If your home is powered by nuclear energy, it is what it is. And most of us aren't producing nuclear weapons or planning where to permanently store waste from nuclear power plants. But if you are on the planning committee for designing permanent nuclear storage, good luck.

Chapter 14

E-WASTE

lectronics produce waste while they're in use and beyond. Aside from the obvious physical electronic objects you use every day, there's also the internet to consider. The internet is a complicated, mysterious, infinite place that is a bizarre hybrid of physical and invisible technology. Most of us probably envision the internet as being relatively contained to our computers and phones—our email goes to someone else's computer, their response comes back to us, and so on. Maybe a satellite or phone tower is involved in the mix as well (they are). Or maybe it lives in an invisible nebula floating above us all ("the cloud"); but those emails, photos, texts, and cloud documents have to live in a physical space—data servers. And data servers are often large trash bins for what is called "digital waste."

Every email sitting in your inbox is housed in one of the almost eight thousand data centers somewhere in the world. Data centers use 1 to 1.5 percent of the world's electricity and create 0.3 percent of global carbon emissions—the same amount as the airline industry—even as less than 40 percent of that energy is used to actively send and receive data.[1] Each center is constantly cooled using enormous amounts of water to avoid the machines overheating, and most facilities run at full capacity 24/7 in the event a surge of data activity happens. Many of the largest data centers are in hot climates that pump potable water from parched communities—a poor choice for facilities that need to be cool at all times. Your family iCloud photo album of a bunch of relatives you haven't spoken to in a decade and never even liked in the first place is being continually cooled in perpetuity just by virtue of its existence online, even if no one has looked at it in years or ever will again.

TRASHY TIDBIT
September 22, 1983:
Atari Video Game Burial, Alamogordo, New Mexico

An urban legend about the collectively agreed–upon worst video game ever made, Atari's *E.T. the Extra Terrestrial*, was discovered to be kind of true in 2015. Video game fans long believed that the much–berated E.T. game was buried and cemented in by the thousands in the desert of New Mexico after its epic failure to sell well in 1983. In 2015, the site of the supposed mass burial was dug up, re–vealing about 1,300 E.T. games and 728,000 other Atari titles.

In 2021, there were an estimated 319.6 billion emails sent and received per day across the globe.[2] Our data creation and consumption is grow-ing at lightning pace as we increasingly rely on internet-enabled technol-ogy. Every email sent, video streamed, game played, question Googled, tweet tweeted, or photo uploaded to the cloud lives in a physical building. These "hyperscale" buildings are *huge*. One very self-importantly named data center, the Citadel Campus in Reno, Nevada, is 7.2 million square feet—compare that to the Mall of America at 5.6 million square feet—and that's actually the *second*-largest server, beat out by a 10.7-million-square-foot center in China.[3] Amazon, Google, and Microsoft own an enormous amount of data servers, renting out space to other huge companies like Netflix. Many of these data centers are powered by fossil fuels, particu-

NO DATA CENTER IN SEALAND

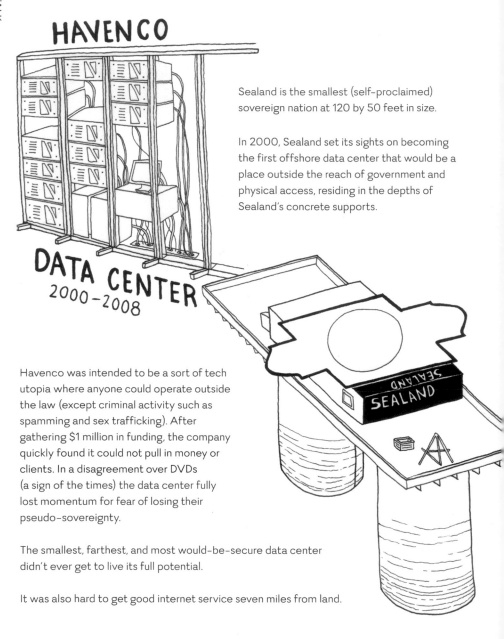

HAVENCO

DATA CENTER
2000-2008

Sealand is the smallest (self-proclaimed) sovereign nation at 120 by 50 feet in size.

In 2000, Sealand set its sights on becoming the first offshore data center that would be a place outside the reach of government and physical access, residing in the depths of Sealand's concrete supports.

Havenco was intended to be a sort of tech utopia where anyone could operate outside the law (except criminal activity such as spamming and sex trafficking). After gathering $1 million in funding, the company quickly found it could not pull in money or clients. In a disagreement over DVDs (a sign of the times) the data center fully lost momentum for fear of losing their pseudo-sovereignty.

The smallest, farthest, and most would-be-secure data center didn't ever get to live its full potential.

It was also hard to get good internet service seven miles from land.

larly coal, and while many claim they are transitioning to becoming carbon neutral, their consumption is projected to grow as more centers are built to keep up with demand.

Technology is becoming more ubiquitous inside the home and on our bodies, and the invisible waste we produce from those devices is quickly adding up. Video streaming and gaming are particularly heavy hitters, requiring tons of data to be stored on servers and accessed for long periods of time. "Cloud gaming" uses software run from a remote data center, rather than on a local home console, and produces tons of digital waste.

The Crypto Problem

One of the biggest culprits of digital waste is cryptocurrency. I'm going to be honest here: I don't totally understand cryptocurrency. I've done plenty of research over the years and learned as much as I can, but struggling to understand the basics of how the internet works on a fundamental level makes it hard to grasp such an abstract use of the system. The internet's building blocks are a realm of information I can't totally wrap my head around—how just 0s and 1s can build something simultaneously so small and so infinite. With that being said, I am happy to say that it doesn't actually matter if I can explain the intricate ins and outs of how the crypto world utilizes the internet, in order to explain the ill effects it has as a byproduct of that process.

When Bitcoin first came on the scene in 2009, it was possible to "mine" for Bitcoins on a home computer, as the competition was not very robust at the time. Now it has expanded into an enormous global market. Alex de Vries, a data scientist who created the Bitcoin Energy Consumption Index, describes mining as "three million machines around the world participating in a massive game of 'guess the number,' generating 140 quintillion guesses every second of the day, nonstop." A quintillion has eighteen zeros: that's 140,000,000,000,000,000,000 guesses a second.

Specialized machines are used in Bitcoin mining that are updated every 1.5 years on average to increase speed and efficiency. These outdated machines, which are laden with toxic materials, are rendered useless and disposed of, contributing up to 33,800 tons of e-waste annually.[4] Mining uses a huge amount of server space that is constantly operating at full capacity, but in its wake leaves an immense amount of wasted electricity. As of 2021, mining Bitcoin uses more energy per year than the entire country of Argentina or Finland. To put it into a more digital sphere, Bitcoin mining uses seven times more electricity than Google.

NUMBER OF MATERIALS USED TO MAKE:

240

300+

120+

130+

Hardware

Throughout our transition from a paper-based society full of newspapers, copy machines, and physical mail into a primarily digital world, we have yet to figure out a sustainable way to support the world's digital use. The internet is strongly tied to waste generation in the form of both physical hardware and electricity consumption. The scope of technology has expanded into every facet of our daily lives and is disposed of at a shocking rate, to the tune of more than fifty-seven million tons of electronic waste per year, less than 20 percent of which is recycled. The UN reports an expected doubling of e-waste by 2050 if nothing improves, with rates increasing by two million tons per year.[5] Many other countries in the world surpass the United States in the effort to *actually* recycle their electronics, such as Europe's 35 percent recycling rate for e-waste. It follows suit that the United States lags significantly behind much of the world's increasing commitment to deal with the waste they produce.

Like much of the world's waste, e-waste is frequently shipped to poor countries to be hand sorted—an extremely dangerous job that leaves the people doing it with essentially no compensation for their work. Until 2017, China received 70 percent of the world's e-waste, at which time they imposed a ban to limit the amount of imported waste. Because of that, the waste previously sent to China is now funneled to smaller, poorer countries. Much of the time workers do not wear protective gear when sorting or breaking down materials and extracting elements that are extremely dangerous, leading them to come into contact with carcinogenic or other hazardous substances.

In the United States, some of the country's e-waste processing has been assigned to prisoners, fitting into a long history of assigning prison labor to dangerous, grueling, or undesirable work for little to no money (often less than one dollar per hour). Prisoner labor in the United States is slavery, as workers are not given rights under the Thirteenth Amendment, which protects against involuntary servitude. They are not granted basic worker protections or rights, and much of the time are not regulated by OSHA standards of safety. Federal Prison Industries, Inc., doing business as UNICOR, is the department that assigns employment to prisoners. In 1997, as e-waste was beginning to increase, UNICOR began accepting devices to be recycled at prison facilities. After an extensive investigation into the condition of these e-waste recycling sites, the US Department of Justice found egregious health and safety issues that were endangering the workers with exposure to toxic chemicals and metals, particularly cadmium and lead.[6]

Outside the United States, e-waste recycling and dumping is an even more severe issue. Out of the 186 countries that signed on to the Basel

Convention—a 1989 international treaty created to stop the international movement of waste—the US is the only country that has not ratified it, enabling the continued export of e-waste. Had they signed on, it would make the export of e-waste illegal and force the country to create federal recycling programs—a more costly endeavor than shipping it to places such as Hong Kong, Ghana, India, and Pakistan. Before China's import ban, Guiyu, in the Guangdong province, received huge amounts of electronic waste from within the country and internationally. The soil in Guiyu has the highest rate of carcinogens in the world due to electronic waste burning.[7]

E-waste is costly and dangerous to recycle, however, the elements contained in the waste are worth an enormous amount of money and have the potential to reduce reliance on virgin mined materials. A total of 7 percent of the world's gold is locked up in e-waste, of which 80 percent is sitting in landfills. If all the gold and silver were to be extracted from America's thrown away phones, it would be worth $60 million a year.[8] This is perhaps the best example of the cradle-to-grave pipeline, with an industry performing the costly process of mining precious materials to create products that will be obsolete in less than a decade. These extremely hazardous materials get tossed into landfills and become very difficult to extract and therefore reuse. Truly, the economic logic behind e-waste is a telling portrait of our lack of care for the planet's resources.

A circular tech economy is a distant dream requiring a seismic shift in production, material generation, and economic mentality, but it is possible. Companies are beginning to accept returns of their products to reuse the materials in a fashion that resembles more of a closed loop. They could also adopt more eco-friendly designs that use materials less detrimental to the environment and human health. And maybe next time you are finished with an electronic device, look into local e-cycling programs through big-box stores like Best Buy, your municipal waste facilities, or nonprofit organizations that refurbish electronics. Or maybe just don't buy the latest phone if yours still works.

Chapter 15

MEDICAL WASTE

ossed-out gloves, expired medication, and single-use face masks: it adds up. Unlike plastic bags or takeout containers, hospitals probably aren't at the forefront of our minds when it comes to waste generation. But if you've been at a medical facility, you most likely noticed the enormous amount of plastic and single-use items for even basic appointments, let alone surgical procedures.

When a Global Health Crisis Creates a Global Garbage Crisis

In a normal setting, we see the gloves and tongue depressors go into the trash, but the COVID-19 pandemic has not only been a cause of fear, grief, global turmoil, and death; it's also increased our garbage problem.

The main ways to reduce one's chances of contracting COVID-19 have been waste producing—masks, gloves, syringes for vaccines, expired or spoiled vaccines, gowns, hand sanitizer bottles, home test kits, and other personal protective equipment (PPE). The World Health Organization reported a tenfold increase of medical waste directly produced from treating and preventing the illness.[1] By November 2021, it was estimated that eighty-seven thousand tons of PPE waste was generated[2] and approximately three million single-use masks were thrown away *every minute*.[3] I wear reusable, washable masks, in part because I find the disposable masks difficult to fit onto my face, and in part because, at the time of buying them, it was nearly impossible to track down N95s. The supply and demand of PPE during the height of the COVID-19 pandemic made it challenging to make decisions about how to balance waste and health when cloth masks were still relatively unsubstantial early on, medical-grade N95 masks were sold out, and standard, single-use blue masks were what was most widely accessible. If you forgot a mask going into a store, much of the time the blue ones were what was available. Buying a hundred throw-away masks for the same price as fifteen N95s that were reusable still had great appeal, particularly because they were more readily available.

While we were trying to remain protected from COVID-19 in public spaces through mask wearing, we were testing at home to see if we had successfully avoided it. In January 2022, it was estimated that five to six million rapid home tests were taken per day in the United States.[4] We've probably all taken at least one, if not dozens, of rapid tests at home over the last three years, as well as PCR tests at public testing sites. The test kits are little boxes of trash; they come in a cardboard box with paper instructions. While the design slightly varies between brands of tests, the general setup is the plastic rectangle part with the anxiety-producing window that shows you the results (hopefully negative) and a plastic tube of liquid that you dip the swab into, all of which are in sterile packaging. These tests are used for about fifteen minutes and then discarded. If you do have COVID-19 or are anxiously praying that you don't before a party, you might take one as reassurance or take them daily to count down until you can leave your room. The cost benefit of not spreading the disease while producing waste is a good example of when one small, discarded product gets multiplied by millions every day to make for a huge problem.

As of April 24, 2023, there have been more than thirteen billion vaccines administered—an incredible feat to protect billions of people from becoming fatally ill.[5] At the same time, these vaccines produce waste that is potentially harmful during its disposal process.[6] The benefit of accessing potentially lifesaving tests is huge for the health of oneself and those around, but medical waste is still globally impactful, even when it's for important causes. Because COVID-19 vaccines are packaged in multidose vials that are not stable once opened, millions of unused vaccines have gone into the trash, with some pharmacy distributors wasting upward of 45 percent of their inventory. The first vaccine I received was very early on in their distribution. I arrived to CVS at eight o'clock in the evening to wait in line for vaccines that were going to be discarded if not used by the end of the day. I, along with a group of other young nonqualifying people, got the soon-to-be-trashed vaccine I would have had to otherwise wait another month to get. As a work-from-home, alone, completely nonessential worker, I was last in line for the vaccine.

Discarded vials are a very small part of the problem. Many healthcare and waste management facilities around the world were not prepared for such an influx of biomedical waste, creating a situation where hospital garbage was not properly disposed of, and exposing workers to hazardous situations, including needle sticks, pathogens, and contamination from incineration. Because of early fears that COVID-19 was transmitted through surfaces (why people were disinfecting their cereal boxes or handling everything using gloves), much of the waste generated from COVID-related treatment was unnecessarily treated as hazardous. With the combination of waste and recycling facilities being temporarily shut down in the height of the pandemic, and the excessive amount of waste produced from PPE and online ordering, treatment of garbage went haywire around the world. In areas where recycling is processed via hand sorting, in-person work ceased to happen during the most contagious era of COVID-19, leading to burning or piling up potentially recyclable materials that were incorrectly categorized as hazardous.

It's a hard pill to swallow that a pandemic, which, according to the World Health Organization, has caused the death of almost seven million people as of May 2023, has caused such an influx of waste production. Sometimes when it comes to garbage, it feels like one solution causes another problem. The avenue to saving human lives comes at the cost of hurting the planet, and therefore endangering the lives of all species.

Everyday Medical Waste

Unfortunately, the bad news of medical waste isn't isolated to just a global pandemic. For instance, the average US hospital patient generates 33.8 pounds of waste per day.[7] I recently had sinus surgery that took one hour with two hours of recovery. Without insurance it would have cost $15,561.15, and when I looked at the bill, much of it was single-use materials, single pills, IVs, or brand-new bottles of nasal spray that I took two squirts of before tossing.

I recognize much of this is because I live in the US with our messed up healthcare system, where in 2012 an estimated $765 billion was wasted on unused or unnecessarily discarded medical waste.[8] According to a report by the National Academy of Medicine, "the annual waste . . . could have paid for the insurance coverage of 150 million American workers." We have warehouses full of medical supplies that will be landfilled because they have exceeded their arbitrary expiration dates, even though there are many countries that lack adequate supplies. Purchasing new materials to replace the perfectly good ones that are discarded, as well as disposing of the unnecessary waste, costs an inordinate amount of money for the healthcare system, the burden of which falls heavily upon taxpayers.

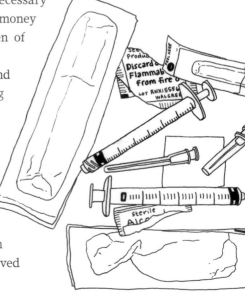

Medical waste is increasing around the world, as some countries have aging populations that require greater levels of care; others are experiencing economic growth, and therefore, greater access to advanced medical facilities.[9] Daily functions are energy intensive, using greenhouse gas–emitting anesthesia, running tons of machinery 24/7, and incinerating much of their hazardous waste. Food is served with disposable cutlery, and much of it is refused or unfinished by patients, while gloves are used in excess in place of handwashing.

Proper disposal of medical waste protects the workers both at the hospital and those receiving the waste, whether that's at a landfill, incinerator, or unregulated dump. But when almost all waste from an operating room, such as unsoiled gloves, plastic wrappers, gowns, and sterilization wrapping goes into the regulated medical waste (RMW) stream, it significantly raises costs and environmental impact. RMW costs ten to twenty-five times more to dispose of than non-biohazardous waste, and yet many hospitals and medical facilities mis-categorize huge portions of their waste as hazardous. Actual RMW is generally defined as materials that have the potential to transmit infectious disease, including but not limited to body parts, bloody bandages, lab cultures, used needles, and some surgical instruments, though there is one point of contention and confusion: "free-flowing blood."

If you were to look at a lineup of objects that have blood in some state on them—a freshly severed arm, a rag caked with dry blood, an IV tube with kind-of-wet/kind-of-dry blood, and a sponge that was used but not that recently—which would you say has free-flowing blood on it? I'd probably say only the arm, but this kind of vague assessment has created a world of confusion and category chaos, leading to facilities going overboard, tossing things that could even slightly, maybe, possibly be thought of as potentially infectious into the biohazard bag to avoid risk of legal ramifications. Legitimate regulated medical waste is a very small portion of the total medical waste from hospitals, but more often than not there is a serious lack of education and sorting infrastructure in place in medical facilities.

There's been an interesting conversation happening in the medical field around the Hippocratic oath that all doctors take to "do no harm." To what degree does that oath extend? Some argue that it ought to extend past the hospital to local, national, and global levels when it comes to contributing to the production of toxic environments that cause increased medical issues for residents living near medical disposal sites. Incinerated RMW can produce dioxins and heavy metals, both of which are extremely harmful to human health for those living near the facilities. According to Health Care Without Harm, a nonprofit working to educate and develop sustainable medical waste models, the healthcare industry's greenhouse gas emissions are so high that if it were a country it would be the fifth-largest emitter.[10] The cruel irony of harming people as a by-product of helping others is not lost on some doctors, as there is a push for more sustainable hospital waste management through proper sorting, more environmentally responsible supplies, and an overall reduction of waste generation.

Unused and Trashed

While much of the issue lies in what is improperly thrown away after use, another piece of it is in what's thrown away without being used. Some medical supplies, similar to food labels, have expiration dates, many of which bear no reflection of the product's quality. I have looked up many times whether I can still take a medication that's five years past its printed expiration date. The answer, according to Roy Gerona, a researcher who studies the effectiveness of drugs years or decades after their expiration dates, is: yes. So, why are so many pharmaceuticals tossed by hospitals, pharmacies, stores, and individuals every year? The expiration date is not an indication of when a medication will become harmful or even less effective, but rather it's the date until which the FDA can confirm maintained effectiveness.[11]

Regardless of the test results achieved for each drug, the FDA deems almost all medications "expired" after a couple of years, some of which Gerona tested thirty years later to reveal their effectiveness had diminished very little, with not a single instance of someone being harmed from taking an expired drug recorded in medical literature.[12]

The pharmaceutical-industrial complex is one of the most greedy, corrupt businesses out there, so it makes sense that there would be little desire or incentive to argue for longer windows of shelf life in stores or hospitals. The financial incentive of constant turnover works in their favor. In a truly maddening slap to the face for individuals who cannot access medication, due to short supply or small pharmacies who lose tons of money on expired drugs, the government is happy to ignore its own rules for the sake of saving money. While it requires the disposal of tons and tons of useful, quality medicine every year, the US federal government has been maintaining its own enormous stash of drugs for the last three or more decades, on deck for some yet-to-happen emergency. The Department of Defense, together with the FDA, has a program that grants shelf life extensions to medications in their own stash that do not apply to any entity outside of the government's stockpile. There have been attempts to change the expiration date rules for the general public with no luck.

There are small-scale efforts to redistribute medications by groups like SafeNetRx, an NGO (nongovernmental organization) that recycles and redistributes medications from nursing homes to doctors or clinics that will prescribe them to patients at no cost.[13] In states where it's legal to recycle medications, these programs help patients in need of medications they cannot access, reduce unnecessary waste, and prevent disposing of pills that would otherwise end up in incinerators or pollute groundwater once flushed. Like food waste, diverting these perfectly good medications to those who need them would alleviate human suffering, economic loss, and environmental damage.

It's not just medications that are dumped from hospitals and nursing homes, but tons and tons of perfectly good supplies. Elizabeth McLellan was formerly a nurse and repeatedly saw that "when patients were discharged, hospital staff threw out everything [in their rooms], including unopened supplies."[14] She founded Partners for World Health, a nonprofit that distributes unused medical supplies overseas and to rural hospitals in the US. There is an appeal to single-use sterile products—it takes no time to clean them and prepare them for reuse, they feel inexpensive, and it's simpler to just use and toss. But like much of our feelings about convenience and expense, cheaper options end up being more expensive in the long term, and they produce mountains more waste, particularly if they're never used at all.

Medical care is so wildly inequitable on hyperlocal and global scales. Reducing medical waste should come from reinvesting in reusable products that can be sterilized—in turn cutting healthcare costs—as well as by legalizing and enforcing donations of unused goods that are desperately needed by underfunded facilities or facilities with little access to goods, both domestically and abroad. The Hippocratic oath needs updating to adjust and account for the emerging ways we are being harmed, and to not unnecessarily add to that harm even more.

Chapter 16

HUMAN WASTE, SEWER SYSTEMS, AND MORE

A book about waste can't overlook the twenty-four-hour waste-producing factories that are our human bodies. We directly produce waste in the form of feces and urine, and subsequently use massive amounts of toilet paper, wet wipes, diapers, and water to deal with it. Human waste isn't listed on municipal solid waste charts since it's handled separately, and the stats are difficult to collect because waste is disposed of very differently depending on national, local, or individual practices. Treatment (or lack thereof) can vary wildly from citywide sewer systems that feed water treatment plants, to rural homes that use individual septic tanks, to villages with no treatment at all, to households that use bidets or composting toilets. Some are cultural or municipal standards, and some are individual choices, but on the whole individuals have little choice in the way their waste is handled. It's estimated that 80 percent of wastewater is not treated before disposal, with wealthier countries having higher rates of water treatment, but still topping out at a rate of only 70 percent treated water.[1]

I spoke to Chris Clapp, executive director of the Ocean Sewage Alliance, about the state of global sewage systems and their effects on waterways, both local and worldwide. Because people don't automatically think of human and domesticated animal waste when they think *trash*, and we don't like to discuss the subject (gross factor), it flies under the radar as a quiet yet enormous problem with far-reaching consequences. The problem is multifold—what we consume, where we live, how the waste is treated, and where it ends up. Like many other kinds of waste, from mining to medical sewage, as a form of trash it is at a point of crisis, and it's largely because we want it to be out of sight, out of mind. Clapp emphasized that "particularly here in Western cultures we want to think that we're so civilized that we have no harm on the environment or are doing our best to have as little harm as possible, and it's that disconnect of people from the larger ecosystem and cycle of things that has created trash."[2]

Waste in public areas, namely bodies of water and streets, has historically been a huge source of environmental and public health issues, ranging from being a vector of the Black Death plague in 1347, to contributing to the deaths of 1.5 million children every year from contaminated water in the modern-day Ganges River. The landscape, infrastructure, and financial status of a region often determines how waste is handled. My house is on a septic system, since I live about one minute past the city line that would provide me with city sewage and water access. I'm decidedly not a prime candidate to be on septic, as I have been actively avoiding calling the septic company for a checkup for two years. I much preferred being on the city's system where there is some comfort in knowing it's being managed by a party outside of myself, and something breaking isn't fully the result of my neglect. However, I can recognize I am lucky to have a clean, accessible way of disposing of my waste, which for much of the world is a huge luxury.

Part of the issue is the natural waste itself, and part of it is how we opt to deal with it. Unlike garbage such as plastics or cardboard, the human and animal waste issue is one that cannot be reduced on the production end—what we make is what we make by being alive, and our global population is quickly growing. So, with that being immutable, the areas of change all revolve around how we decide to treat that waste. There are enormous cultural differences in bathroom etiquette and norms from country to country, and oftentimes regions within them. Bidets are ubiquitous in Japan and Italy; there are separate wastebaskets for toilet paper in Mexico, jars of water in India, and lots and lots of toilet paper and wet wipes in the United States and England. Only about 30 percent of the world regularly uses toilet paper, but that 30 percent is using a huge quantity.

TRASHY TIDBIT

August 8, 2004:
Dave Matthews Band Dumps Sewage, Chicago, Illinois

A tour bus belonging to the Dave Matthews Band dumped eight hundred pounds of the bus's sewage off a bridge and onto a tour boat on the Chicago River. About eighty passengers on *Chicago's Little Lady* were doused with a slurry of sewage, with five passengers taking a detour to the hospital for testing. Everyone on the boat got a refund and the boat began a new tour two hours later.

Fatbergs

If you live in England, home of particularly old, craggy sewage systems, there's a good chance that there are masses growing and lurking in the sewer systems beneath you. Rats snacking on crumbs seem like minor sewer dwellers in comparison to the fascinating, disgusting monstrosities known as fatbergs. Remember that leftover grease you washed down the sink when washing a pot, or the wet wipe that's advertised as flushable that you flushed down the toilet as instructed? Or the grease from your neighborhood restaurant that may have been dumped at the end of the night? Those are the ingredients that are forming shocking subterranean masses.

FOGs (fat, oil, and grease) and non-degrading paper or cloth products are fast friends in sewers, congealing to form huge masses of material that harden into bus-size, concrete-like chunks, causing major blockage issues for city infrastructure. Unlike newer, smoother concrete or PVC pipe systems, old pipes or sewer systems tend to have more textured surfaces that cause materials to snag onto the walls and progressively build up over time like a snowball of garbage, which is why England is home to many of the largest fatbergs on record. Fatbergs are the slimy, underground antithesis of our out of sight, out of mind mentality that our waste just goes away once we discard it. Fatberg-building wastes—cooking oils, food scraps, wipes, bandages, sanitary products, and condoms—leave our homes or restaurants and cause invisible mayhem right below our feet.

GARBAGE DISPOSAL
DON'TS

try to make DIY peanut butter

shove a pile of glutenous or starchy foods down

dispose of the evidence of stringly old (or new) celery

use a fork

reach for stuff

pour boiling water while it's running

dump grease and oils
(hot water loosens these, coating the stuff around)

Wet Wipes

Thanks in large part to misleading corporate advertising and a lack of education about waste systems, our ability to make the right choice is skewed. Wet wipes in particular are a huge source of waste generation. The UK alone uses an estimated eleven billion per year, which increased over the course of COVID-19 due to use of disinfecting wipes and a proliferation of wipes being advertised for adults, not just for babies.[3] Multiple lawsuits have been filed and government efforts are in place to require accurate advertising that they are not flushable, though there are a handful of companies who have made products that actually disintegrate in the wastewater treatment process. Like in many realms of waste and its disposal, the question arises of how to hold corporations accountable. How is it that the wipes companies are allowed to put a "sewer safe" label on their packaging when we know that they're not? How confusing is that? "Everyone's afraid of saying 'you can't' market that even though we know the cost to society,"[4] Clapp explained, pinpointing the main force behind the continued issue of sewage problems in the United States.

Massive sewer clogs cost cities millions of dollars and can create spills, backups, and broken pipes. Charleston, South Carolina, filed a lawsuit against several major manufacturers and retailers of wet wipes in response to a clog so severe that "divers had to swim down 90 feet through raw sewage into a dark wet well to pull a 12-foot-long mass of wipes from three pumps."[5] So, each time you think about flushing a wet wipe, pretend you're that diver who's going to dive through ninety feet of sewage to pull it out.

One of the largest documented fatbergs is the "monster fatberg" in Whitechapel, London, which, when discovered in 2017, weighed in at 140 tons, stretched 820 feet, and required weeks to break apart using high-powered water jets and pickaxes.[6] In 2018, a portion of the fatberg was displayed at the Museum of London, which attracted tons of visitors curious to see the disgusting and fascinating hardened block of sludge, and in the process gain some awareness of their contribution to the problem at hand.[7]

What Goes In Must Come Out

Much like what we flush is an indicator of what's going on aboveground—what oils are being cooked with, where restaurants are concentrated, or which cities use more wet wipes—individual bodies are indicators of our environment as well, and can illuminate a multitude of external factors. What medications are available and therefore excreted? What heavy metals or toxins are present in the food someone eats, the air they breathe, or the water they drink? What communicable diseases or pathogens are present in the community? The phrase "we are what we eat" is only part of it. Really, it's closer to "we are what we excrete." Waste can illuminate unexpected information about what's going on aboveground, like when "cocaine concentrations in US wastewater treatment plants [rose] during and immediately after the National Football League Super Bowl," or how "in Taiwan, levels of cocaine (and many other illicit drugs) spiked in waste water after six hundred thousand people arrived there for a music festival."[8]

Wastewater-based epidemiology has become a useful tool for gathering area-wide data that would otherwise be challenging to collect. It mostly flew under the radar as a specialty method of research until the COVID-19 pandemic brought some awareness to the public of what our wastewater says about our health; by measuring the level of COVID-19 in wastewater treatment plants, scientists were able to gather data about the prevalence of the disease in their collection areas. Wastewater monitoring is a holistic view of a population, which has its benefits and drawbacks—an unbiased data set that helps the overall picture, but eliminates the ability to target individuals or specific areas.[9] For example, New York City has fourteen wastewater treatment plants to serve the entire city. Pinpointing issues in a specific neighborhood would be challenging, but would still give useful insight into general patterns.

From Feces to Fertilizer

While our bodies are full of chemicals—these can include makeup washed off in the sink or conditioner rinsed down the shower drain—those chemicals do not necessarily impede creative solutions to capitalize on reusing that wastewater. The concept of closing the nutrient and water cycle loop is extremely promising for human waste and wastewater. *Dune* has the right idea. Urine collection and diversion is an efficient way of capturing almost all the nutrients in urine and treating it in such a way that it is safe to reuse in the agricultural sector or landscape sector, so chemical-based fertilizers don't need to be imported, primarily from Russia. This solution would close the loop, creating a highly valuable product from a renewable resource that's currently just treated as disposed waste in a linear, not circular, fashion.

One limiting factor is the economics of manufacturing, which currently

TRASHY TIDBIT

February 28, 1983: The New York City sewer system was tested when an esti-mated one million viewers used (and flushed) their toilets after the series finale of M*A*S*H, sending 6.7 million gallons of water into the system at the same time.[10]

favors chemical manufacturing of fertilizers over the start-up cost of estab-lishing wastewater-to-fertilizer capturing processes. While chemicals and nonorganic substances can be neutralized and utilized, it's imperative to shift the responsibility around waterway pollution away from municipali-ties and onto producers as plastics and chemicals become more and more ubiquitous in our waterways.

There Is No Diaper Genie

Despite the branding on that special trash can, there is no diaper genie. Diapers are thrown into the garbage by the millions every year. The av-erage baby in the United States will go through six thousand diapers in the first two years of its life, and worldwide, three hundred thousand dis-posable diapers are tossed every *minute*, while their annual production consumes 248 million barrels of crude oil.[11] Dis-posable diapers have become popular in many countries, replacing reusable cloth options. Some environmental organiza-tions and individuals are trying to push a return to reusable, but it's not so straight-forward. Cloth diapers involve a large up-front cost to purchase enough to not have to be washing them every half hour, as well as access to convenient laundry (whether that's a home washing machine, a pick-up service, or hand-washing). Many parents value the convenience of disposables, and

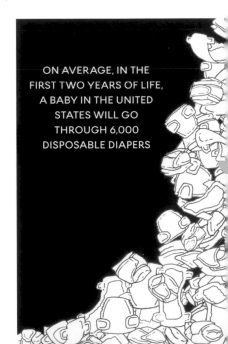

ON AVERAGE, IN THE FIRST TWO YEARS OF LIFE, A BABY IN THE UNITED STATES WILL GO THROUGH 6,000 DISPOSABLE DIAPERS

there are currently no real compostable options that work with municipal or home settings, so using and tossing is the easiest option.

Diapers take over five hundred years to decompose in a landfill, and with so many millions of them occupying landfill space, it seems time for some alternative that is cost-effective (both financially and in terms of time required) and can more easily break down or be reused.

The Dog Poop Dilemma

There is one realm of poop production that is 100 percent consumer responsibility, so listen up, pet owners: it's you. Around 50 percent of US households have at least one dog, which is a lot of dogs—at least ninety million of them. And that means a lot of dog shit to pick up (or leave behind). I've lived with dogs for thirty-four years as of this writing. That means picking up 24,820 to 37,230 poops (I'm going off one dog pooping two to three times a day) if I was the one picking it up at age zero. People have very different opinions about picking it up or leaving it. I do both. When I'm halfway through a walk in the woods and don't want to carry a bag of poop for two miles, I will leave it in the woods and make a lackluster attempt to kick some leaves over it. If I'm in a neighborhood and my dog poops on your lawn, I've got my bag ready.

Some people are inconsistent like me—different situations call for different actions—though after researching this more thoroughly, I am leaning a little more toward one side. In Camp Leave It, the reasons are *It's better to let it be in nature than in a landfill*, *It's one less plastic bag in the garbage*, *It's poop and there are animals everywhere shitting in the woods*, or *I moved it so it's not in the way for anyone to step on*. In Camp Pick It Up, the reasons are *It's gross to leave it*, *It's not a natural substance like other animals'*, *It's polluting*, or, most likely, *It's anxiety-provoking that someone will publicly scold me for leaving it*.

In some part the root of both camps is ethics around which causes more environmental harm, unless it's just laziness or shame. Do I add plastic to the landfill, or do I add poop to the woods? Neither option is great, but I'm here to add information to the debate: You probably shouldn't leave it. Dogs have different guts and different dietary intakes than other animals. We give them medications and feed them diets that aren't reflective of what's available in the environment. Animals that are only eating food directly from their habitats are not going to pose a threat to their environment in the same way that a domesticated animal will. Dogs can introduce pathogens and as much as three times more bacteria than humans to the environment, as well as excess levels of phosphorous and nitrogen that can cause oxygen depletion in water—an environment that's inhospitable for fish and other aquatic life, and a breeding ground for harmful algae.

With about five hundred million poop bags being used globally every year, it seems like that's a pretty bad version of the better option, and maybe, just maybe, those biodegradable bags that are an inexplicably awful powdery texture are better? But alas, we run into the biodegradable plastic issue again. Once in the landfill, these theoretically biodegradable bags of poop essentially mummify without access to oxygen.

There is one small-scale innovation that seems to be effective at keeping it out of the landfill and out of the waterways: a dog-specific mini septic system. A friend of mine discovered this invention in an unpleasant and unexpected way. Unbeknownst to her, buried in her backyard was a doggy septic system that had been sitting there untouched for who knows how long, and in the interest of investigating the mystery contraption, she took off her gardening glove and stuck her bare hand directly into the hole. For the system to work properly, waste is deposited into an underground tank with a lid on the surface, and a treatment of bacteria and enzymes is added to break down the waste, making it safe to leach into the ground. In theory, the waste should become neutralized and enter the soil with no harmful bacteria or pathogens present in dog poop. This system has been installed in public parks in Richland County, South Carolina, as an experiment in sustainability. It can be installed at home with success depending on soil and weather conditions. It's certainly not a perfect, universally accessible solution, but it's a step toward an alternative option that produces less waste than the waste itself.

And if after reading this chapter you're finding yourself curious to spend more time at a wastewater treatment plants, according to Clapp, "they love giving tours, and if they could keep you there, give you dinner, and have you there overnight, and then continue the tour in the morning, they probably would."[12]

Chapter 17

THE FUNERAL INDUSTRY

We make human waste while we're alive, but we also make human waste when we die; our bodies ultimately become waste. We don't like to deal with trash almost as much as we don't like to deal with death. Combine the two and you have a recipe for a very opaque problem that the funeral industry has very little financial interest in making clearer: their garbage and the garbage they sell us.

Until somewhat recently, the legal options for dealing with bodies in the United States were highly regulated, and limited to cremation, burial, or donation to scientific research; and those first two options have serious environmental consequences. The funeral industry is a huge contributor to pollution and waste from sources such as embalming chemicals, cement, and coffins that are more decked out than a typical pine box. Cremation has increased in popularity worldwide, which comes with its own issues of fossil fuel use and greenhouse gas production, ranging from modern crematoriums to India's long-standing practice of wooden pyres, which come with significant waste and pollution issues.

A Specific Case: Indian Death Rituals

Indian death rituals have used wooden pyres for thousands of years. Placing a dead body within a pile of hundreds or thousands of pounds of wood to be set afire is a sacred practice in one of the world's oldest cities, Varanasi, which sits on the Ganges River. Hindus come from all over the country to cremate their loved ones in Varanasi, with the belief that being put into the Ganges River after death is a way to break the cycle of reincarnation. It's estimated that two hundred bodies are burned every day in Varanasi,[1] and one

hundred thousand bodies are put into the Ganges every year.[2] The cremation ghat—a platform on the riverbank where the oven is located and pyres
are set afire—is constantly alight, burning bodies through day and night.
Families usually gather to witness their loved ones' cremations, and it's one
of the biggest expenses for some individuals and families; buying enough
wood to adequately burn the body is not cheap, and is an entire industry
in and of itself, full of illegal logging and underhanded dealings. The ashes
of the dead are then dumped into the river, finishing the sacred mission to
break out of reincarnation.

While this practice isn't generating garbage in the way an old banana
peel or plastic container in a landfill does, it still generates waste. The
fifty to sixty million trees used specifically for cremations release tons
of CO_2 into the atmosphere when burned, and the ashes along with the
bodies' remains are dumped into an already polluted river.[3] Sometimes
bodies are not completely burned and are put into the river still somewhat intact. Members of the government and environmental sectors have
encouraged making changes to this ancient tradition by way of using less
wood or implementing traditional crematoriums, in order to improve the
water quality of the river, reduce deforestation, and improve health for the
surrounding citizens, but no changes have been made on a large scale. To
many Hindus, the river itself symbolizes purity, which isn't dependent on
its quality or cleanliness.[*]

Western Practices

Western burials generate massive amounts of garbage in a more direct
sense of the word (i.e., *stuff*), particularly among those who opt for opu-
lent burial products and services. I have chosen, at the risk of sounding
crass, to consider all nondegradable material goods that are used in burial
practices as garbage, including the chemicals pumped into those who are
embalmed. Things such as nonwooden coffins with metal handles, plastic

---

[*] As someone who does not practice Hinduism and isn't from India, I do not have a personal stance
on the ritual of funeral pyres. Honoring and respecting a culture's long held sacred ritual is important,
particularly for those of us who are completely outside the tradition, and to point out the environmental
issues that stem from the practice is not to imply it should not exist. I understand the basics of how the
process is done and what health effects it has on the river, and can only speak to that information, which
is not informed by a personal relationship to the religion, country, or culture.

casket liners, plastic flowers left for the dead, or concrete used in burial vaults take hundreds or thousands of years to break down, creating garbage the second they are lowered into the ground or left on a grave. As for the person buried, if they were embalmed, they're emitting waste, too.

Nondegradable caskets , including a fish in the style of Ghana's playful casket making

Western funeral traditions became immersed in waste-generating practices in earnest beginning around the American Civil War when chemical embalming came into practice, implementing heavy metals and arsenic to preserve bodies for viewing.[4] Embalming became the cash cow of the funeral industry, capitalizing on the dead's loved ones' desire for open-casket funerals and a version of death that denied the less glamorous, more truthful, part of it: decomposition. Embalming is a wildly wasteful way to deny death.

According to the Green Burial Council, the annual stats surrounding traditional burials are staggering, amounting to 4.3 million gallons of embalming fluid, 20 million board feet of hardwoods, 1.6 million tons of concrete, 17 thousand tons of copper and bronze, and 64,500 tons of steel.[5] The funeral industry profits from our denial of death, taking in a whopping $20 billion annually. Some people opt for concrete burial vaults, which are cement boxes that house the casket underground, preventing the earth above from settling, to preserve a manicured lawn in the cemetery. Putting a casket inside a concrete box is essentially manufacturing an object whose whole purpose is to immediately be buried and never seen again. A casket on average costs between $2,000 and $5,000, with the highest-priced ones fetching over $10,000, while a simple wooden box costs under $500.

AN UNFORTUNATE MISHAP

Sometimes people pay a lot of money to end up accidentally destroying their loved ones' remains. In rare cases, "exploding casket syndrome" occurs—a phenomenon in which a desperate effort to prevent a body from decaying royally backfires. Air exposure isn't the culprit of decomposition, and depriving a body of the outside elements doesn't prevent decay. The body takes care of that itself, beginning to decompose the moment after death (unless it's frozen). In the bloat phase of decomposition, the body begins liquefying and producing gases. We've all seen the Mentos into Coke experiment . . . now imagine that as a body in a casket. While putting a body in a sealed vessel may seem like the way to preserve it, it does quite the opposite.

Consider mummies, the most preserved bodies on Earth. After removal of their internal organs, the bodies were treated with sand, salt, air, and linen, which allowed the bodies to maintain a high level of detail and structural integrity for thousands of years. Removing all potentially putrid, gassy, liquefying internal organs and encouraging desiccation is what preserved a body, not locking it away to digest itself with no opportunity to dry out. If you take the opposite (and modern) approach, rather than getting a dry bag of bones for three thousand years, you get a pressure cooker of guts that are expanding much beyond the allotted space of the casket or mausoleum. So, maybe rethink putting Grandma in a sealed container. Some caskets are now designed with an outlet to release gas buildup and prevent her from oozing out.

A Greener Way to Die

In a rejection of stuff-oriented, death-avoidant burials, there is a growing movement toward green burials—a return to the way most people were buried throughout history until a couple hundred years ago. Some natural burials happen in traditional cemeteries, and others are conservation burials that are part of an effort to restore natural habitats and/or encourage a more thoughtful relationship between death and the natural world. Natural burials are built into some religions and spiritual practices, like Judaism and Islam, which forbid embalming and advise burying the dead in shrouds or simple wooden caskets.

A friend of mine, Katrina Spade, developed Recompose, the first human-composting operation. Bodies are brought to a facility in which loved ones can have a ceremony before the composting process begins, ultimately producing soil that loved ones can take home to use. What is not taken by the families is distributed at Bells Mountain for habitat restoration. This process requires very little energy and produces no physical waste. It is becoming legal in more states throughout the country.

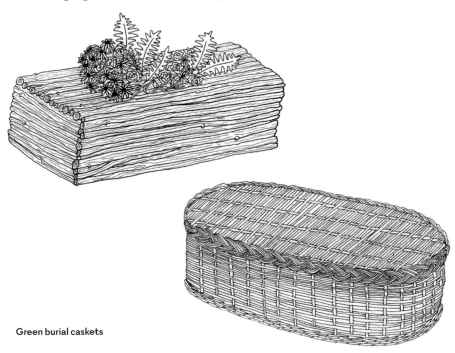

Green burial caskets

Dedicated natural burial sites are becoming more popular around the world, often set in meadows and woodlands where the public can walk or hike throughout the parks. I have spent time at the Larkspur Conservation burial ground in Nashville, Tennessee. The park is a 112-acre conservation easement and land trust working to restore much of its ecosystem to reestablish biological diversity, as well as provide people an alternative to traditional cemetery burials or mausoleums. Natural cloth shrouds, simple wooden caskets, or handwoven baskets are common ways to be buried there and at many other green burial sites; once the body is buried, a stone can mark the grave, and native species, over time, grow on and around the grave. I took a long hike through meadows and forests with a friend who helps with burials, and every so often would walk past graves, some of which just looked like tiny mossy hills, and others a stone along a path through a field. It was beautiful and serene knowing there weren't hundreds of thousands of dollars of plastic and concrete preventing the bodies from becoming part of the landscape. It was a peaceful, relieving feeling that I hope becomes more common as a way to honor that humans are not separate from the natural world, and to reduce the amount of waste generated by many modern funeral practices.

Chapter 18

ORBITAL DEBRIS

The image most of us probably think of when we imagine Earth from outer space is that classic photo, *The Blue Marble*, taken from the Apollo 17 mission in 1972. Floating majestically in a sea of infinite black, Earth sits pristinely against the background. That photo was taken from a space mission to land on the moon. Our image of Earth's pure, unmarred place in the stark darkness of outer space is not quite as accurate these days.

We have been polluting Earth's land and sea for decades, but since the launch of Sputnik 1—the first artificial satellite—in 1957, our trash has expanded to outer space. As humans travel further and more frequently into space, we leave more and more debris in our wake. From the tiniest flecks of paint to large defunct satellites, our orbital litter is beginning to pose real problems for the health of space and Earth as it becomes a rotating landfill above us. I believe outer space is just as entitled to its own health and well-being, even if it's not a living, breathing ecosystem like ours. It has become a tragedy of the commons: an unregulated public resource that becomes overused, depleted, or exploited due to individuals, corporations, or governments acting in their own self-interests, to the detriment of the common good. Earth's orbit is now home to national space missions from dozens of countries and satellites owned by the private sector, which contribute to the fastest-growing increase in satellite traffic.

Private companies are quickly introducing "megaconstellations": fleets of communication satellites in low orbit, far outnumbering scientific or research vessels and posing a risk of congestion and collision. SpaceX, the space exploration and technology company started by Elon Musk, has

developed a high-speed internet service called Starlink and is sending up satellites by the thousands. As of 2023, Starlink has launched five thousand satellites, has approval for eleven thousand more, and has requested an additional thirty thousand from the Federal Communications Commission. An estimated 10 percent of Starlink satellites in orbit will soon be nonoperational, beginning a slow process of planned removal from orbit, and sending tons of material into the atmosphere every day as they fall from space.[1] Because Starlink dominates Earth's lower orbit, it's also the culprit of most near misses with other orbiting objects, heading toward a rate of 90 percent of all near misses.[2]

While some orbital debris is programmed to return to Earth's atmosphere once defunct, some of this dumping is intentionally planned to stay in space, and some is a by-product of missions that have returned. The various missions to the moon have left behind over four hundred thousand pounds of trash on its surface, and the amount of garbage orbiting Earth is staggering. According to NASA, there are more than twenty-three thousand pieces of junk greater than a softball, five hundred thousand pieces greater than a marble, and over one hundred million pieces approximately one millimeter in size.[3]

From tiny fragments of material to bigger remnants crashing into other junk, it all can cause major problems for active equipment orbiting Earth. Even the smallest pieces can have damaging effects at speeds of around 15,700 miles per hour. The impact of these objects can cause items to fall back to Earth or break satellites, which in turn become more defunct space junk. For instance, in 2007, a Chinese missile was sent to intentionally destroy an old satellite and resulted in over 3,500 new pieces of orbital debris.[4] This debris will eventually hit other objects that will break apart and hit other objects that will hit other objects, and the process will continue indefinitely, exacerbating the issue and limiting the potential to clean it up.

Along with the potential (and reality) of debris causing damage to other orbiting objects, scientific research is being jeopardized—satellites in orbit are at risk for physical damage, and both satellite and ground research experience light pollution that impedes exploration. Attempts are being made to address the issue and implement cleanup strategies, ranging from sending defunct objects into higher orbits to declutter the active orbit zone, to developing satellites that will latch onto objects, removing them from independent orbit. While these are useful ideas and ones that can be put into place, the scale of debris is worsening, and the technology to clean it up is not matching the pace of debris creation. Not to mention, the smaller the debris is, the harder it is to collect. Imagine trying to pull a fleck of paint out of orbit. It's a similar undertaking to ridding the ocean of microplastics.

Not all the garbage we've put in space is orbiting—some of it is littering Mars and the moon. There are currently almost sixteen thousand pounds of dead equipment and remnants of crashed spacecrafts on the surface of Mars, and most if not all will remain there until they break down, which will likely never happen.[5] Unlike the risks posed for communication satellites in orbit, Mars's litter doesn't negatively affect any daily functions on Earth. But it's a glaring representation of humanity's desire to conquer and explore exceeding our desire to respect the spaces we occupy, even if those uninhabited spaces are 53.4 million miles away. Similar to how we treat the ocean, we regard the places we can't see or use as fair game to be dumping grounds. Outer space is an untouched, mysterious place, and it seems to me that humans want ownership of it, even if it comes at the price of tarnishing what's not ours to take. Just because no humans are there doesn't make it ours.

When it comes to the moon, it's no surprise that on missions to essentially conquer it, astronauts from the many countries that have landed manned or unmanned missions to the moon have left loads of trash there (as of 2022, eight countries have littered on the moon).[6]

A Partial List of Items on the Moon[7]

- Five American flags (seems like overkill)

- Ninety-six bags of vomit and human waste

- Three rovers

- Various tools

- Improvised javelins

- A family photo

- Two golf balls

- A disk containing messages from the leaders of seventy-four countries

- The video camera that captured the famous video of astronauts on the moon's surface

Some of this is equipment that was discarded so moon rocks could be brought back on the ship; some of it was purely in the name of fun, such as golf clubs; and some of it was sentimental, like family photos. Leaving flags or bags of vomit has the same end result, even if one seems explicitly patriotic—it is symbolic of the desire to colonize what we feel entitled to take. This evidence will remain in place forever until it is bleached to an unrecognizable state by UV rays, taken by aliens, or exploded in a cosmic collision.

CONCLUSION

As you come back down to Earth from floating up with the orbiting paint flecks and Starlink satellites, you might put this book down and look around the room, noticing plastic galore, lithium batteries, mined metals, and polyester clothes. Or if you're outside, you may notice the electric cars or trash bags piled into a dumpster. I hate to tell you, but you might know too much now to unsee the information you've learned. It's a distressing and distracting thing, but also a good one.

As we get closer and closer to catastrophe with climate change, there will at some point need to be a forced awakening around the global system of waste. I recognize that *catastrophe* may seem like an extreme classification, but I assume you might feel like it's justified, now that you know more about the garbage crisis at hand. We need to deal with our trash by producing less of it for the sake of the environment we inhabit and the less privileged populations that have to deal with it.

Having context for any issue is imperative to understanding how we fit into it, knowing what changes or contributions are realistic as individuals (who aren't the CEO of Coca-Cola or Jeff Bezos), and having greater awareness of the lives and well-being of people around us, far and near. As stated in the beginning of this book, this is not a guide to *do* better, but one to *understand* better. Whether that's knowing where your clothes end up after only six months of wearing them, or how much food you might be throwing out per year without realizing, a shift in awareness is as important as a shift in behavior. It creates empathy and insight around privileges and hardships in the global garbage issue and allows us to see elements of this problem that we might not have even known existed before.

You don't need to go out and join Greta Thunberg or never buy plastic again—don't set yourself up for failure or take a deep dive into the world of your own eco-guilt—but next time you want to flush that wet wipe, think of the poor guy whose job it will be to fish it out of the fatberg. Making daily choices might not save the world, but you can at least make someone's life a little less hellish until we're given better choices by those up top. And that makes a difference.

ACKNOWLEDGMENTS

A professional thank-you to Bryant Holsenbeck, Cassie, Chris Clapp, Hunt Revell, Irene Li, Kate Woodrow, Lauren Appleton, Maggie Lee, Mei Li, and Sam Hawes, and a personal thank-you to Cassy, Gabriel, and Rachel.

ABOUT THE AUTHOR

Iris Gottlieb is an author and illustrator working to make information more accessible and interesting to a wide range of readers. Iris is the author of *Seeing Science*, *Seeing Gender*, *Natural Attraction*, and *Everything Is Temporary*, and during nonworking hours is poorly playing basketball, making miniatures, and researching sharks.

NOTES

Introduction: So, What Is Trash?

1. Goldstein, Nora, Rob van Haaren, and Nickolas Themelis. "The State of Garbage in America." *BioCycle* 51, no. 10 (October 2010): 16. www.biocycle.net/the-state-of-garbage-in-america-4.

2. PBS. "NOVA Online Adventure: Scaling the Pyramids." Accessed September 9, 2022. www.pbs.org /wgbh/nova/pyramid/geometry/index.html.

3. Lohuizen, Kadir van. "Drowning in Garbage." *The Washington Post*, November 21, 2017. Video, 1:57. www.washingtonpost.com/graphics/2017/world/global-waste.

4. Strasser, Susan. *Waste and Want: A Social History of Trash.* New York: Metropolitan Books, 1999.

5. Hickey, Laura S., and Kristy M. Jones. "A Research Study on Textbook Recycling in America." *National Wildlife Federation*, November 2012. www.nwf.org/~/media/PDFs/Eco-schools/McGraw%20 Hill/12-4-12%20A%20Research%20Study%20on%20Textbook%20Recycling.ashx.

6. Bookshop.org. "About Us." Accessed November 19, 2022. www.bookshop.org/info/about-us.

7. Bookshop.org. "About Us." Accessed November 19, 2022. www.bookshop.org/info/about-us.

8. Leonard, Annie. *The Story of Stuff: How Our Obsession with Stuff Is Trashing the Planet, Our Communities, and Our Health—and a Vision for Change.* New York: Free Press, 2010.

9. Garber, Megan. "It Takes More Than 3 Gallons of Water to Make a Single Sheet of Paper." *The Atlantic*, July 17, 2013. www.theatlantic.com/technology/archive/2012/06/it-takes-more-than -3-gallons-of-water-to-make-a-single-sheet-of-paper/258838.

10. Leonard, Annie. *The Story of Stuff: How Our Obsession with Stuff Is Trashing the Planet, Our Communities, and Our Health—and a Vision for Change.* New York: Free Press, 2010.

11. World Intellectual Property Organization. *The Global Publishing Industry in 2018.* Geneva: World Intellectual Property Organization, 2020. www.wipo.int/edocs/pubdocs/en/wipo_pub_1064_2019.pdf.

12. Penguin Random House. *Social Impact Report.* New York: Penguin Random House, 2020. social -impact.penguinrandomhouse.com/wp-content/uploads/2021/05/2020-Penguin-Random-House -Social-Impact-Report.pdf.

13. Rainforest Action Network. *Beyond Paper Promises: Assessing the Impacts of Corporate Pulp and Paper Commitments on Forests and Frontline Communities.* San Francisco: Rainforest Action Network, 2018. www.ran.org/wp-content/uploads/2018/06/RAN_Beyond_Paper_Promises_2018.pdf.

14. Hachette UK. "About Us: Sustainability and ethics." Accessed January 9, 2020. www.hachette.co.uk /landing-page/hachette/hachette-and-the-environment.

15. Goleman, Daniel and Gregory Norris. "How Green Is My iPad?" *The New York Times*, April 4, 2010. archive.nytimes.com/www.nytimes.com/interactive/2010/04/04/opinion/04opchart.html.

Chapter 1: Ancient Systems

1. Rathje, William, and Cullen Murphy. *Rubbish!: The Archaeology of Garbage.* Tucson: The University of Arizona Press, 2001.

2. Canadian Museum of History. "Archaeological Excavation: Shell Middens." Accessed September 10, 2022. www.historymuseum.ca/cmc/exhibitions/aborig/tsimsian/arcmidde.html.

3. Lubofsky, Evan. "In the Chesapeake Bay, Shell Mounds Show a Long History of Sustainable Oyster Harvests." *Hakai Magazine*. November 22, 2018. https://hakaimagazine.com/news/in-the -chesapeake-bay-shell-mounds-show-a-long-history-of-sustainable-oyster-harvests/.

4. Hunt Revell, in discussion with the author, September 2022.

5. Klivans, Laura. "There Were Once More Than 425 Shellmounds in the Bay Area. Where Did They Go?" KQED, March 24, 2022. www.kqed.org/news/11704679/there-were-once-more-than-425 -shellmounds-in-the-bay-area-where-did-they-go.

6. The Historical Museum of Jomon Village Oku-Matsushima. *The Seaside Village of the Jomon Period: National Historical Site, Satohama Shell Mounds*. Japan: The Historical Museum of Jomon Village Oku-Matsushima. Accessed September 10, 2022.

7. Hunt Revell, in discussion with the author, September 2022.

8. Hunt Revell, in discussion with the author, September 2022.

9. Rathje, William, and Cullen Murphy. *Rubbish!: The Archaeology of Garbage*. Tucson: The University of Arizona Press, 2001.

10. Hunt Revell, in discussion with the author, September 2022.

11. Gannon, Megan. "1,500-year-old Garbage Dumps Reveal City's Surprising Collapse." *National Geographic,* March 25, 2019. www.nationalgeographic.com/culture/article/ancient-garbage-dump -elusa-reveals-surprising-city-collapse.

12. Squires, Nick. "Roman Rubbish Dump Reveals Secrets of Ancient Trading Networks." *The Telegraph,* June 4, 2015. www.telegraph.co.uk/news/worldnews/europe/italy/11650703/Roman -rubbish-dump-reveals-secrets-of-ancient-trading-networks.html.

13. Bar-Oz, Guy, Lior Weissbrod, Tali Erickson-Gini, Yotam Tepper, Dan Malkinson, Mordechay Benzaquen, Dafna Langgut, et al. "Ancient Trash Mounds Unravel Urban Collapse a Century before the End of Byzantine Hegemony in the Southern Levant." *Proceedings of the National Academy of Sciences* 116, no. 17 (March 2019): 8239–48. https://doi.org/10.1073/pnas.1900233116.

14. Alberge, Dalya. "Pompeii Ruins Show That the Romans Invented Recycling." *The Guardian,* April 26, 2020. www.theguardian.com/science/2020/apr/26/pompeii-ruins-show-that-the-romans -invented-recycling.

15. Alberge, Dalya. "Pompeii Ruins Show That the Romans Invented Recycling." *The Guardian,* April 26, 2020. www.theguardian.com/science/2020/apr/26/pompeii-ruins-show-that-the-romans-invented -recycling.

Chapter 2: So, How Did We Get Here from Pompeii?

1. Guldager Bilde, Pia, and Søren Handberg. "Ancient Repairs on Pottery from Olbia Pontica." *American Journal of Archaeology* 116, no. 3 (2012): 461–81. https://doi.org/10.3764/aja.116.3.0461.

2. Strasser, Susan. *Waste and Want: A Social History of Trash*. New York: Henry Holt and Company, 2000.

3. CBS News. "Garment Workers in Los Angeles Describe the 'Modern-day Slavery' of Sweatshops: 'They Paid Us Like 5 and 6 Cents for a Piece.'" CBS News, September 14, 2021. www.cbsnews.com /news/la-garment-factories-investigation.

4. Ratcliffe, Rebecca. "Major Western Brands Pay Indian Garment Workers 11p an Hour." *The Guardian,* February 1, 2019. www.theguardian.com/global-development/2019/feb/01/major -western-brands-pay-indian-garment-workers-11p-an-hour.

5. Thompson, Eliza. "Justin Bieber Says He Never Wears the Same Underwear Twice." *Cosmopolitan,* May 21, 2015. www.cosmopolitan.com/entertainment/celebs/news/a40821/justin-bieber-james -corden-car-karaoke.

6. MacKinnon, J.B. "The L.E.D. Quandary: Why There's No Such Thing as 'Built to Last.'" *The New*

Yorker, July 14, 2016. www.newyorker.com/business/currency/the-l-e-d-quandary-why-theres-no
-such-thing-as-built-to-last.

7. NPR. "The Phoebus Cartel." *Throughline,* March 28, 2019. Podcast MP3, 32:00. www.npr
.org/2019/03/27/707188193/the-phoebus-cartel.

8. Gonen, Ron. *The Waste-Free World: How the Circular Economy Will Take Less, Make More, and Save the Planet.* New York: Portfolio / Penguin, 2021.

9. History.com Editors. "Ford Motor Company Unveils the Model T." HISTORY, July 28, 2019. www.history.com/this-day-in-history/ford-motor-company-unveils-the-model-t.

10. United States Census Bureau. "US Census Bureau History: Ford Model T." United States Census Bureau, October 2018. www.census.gov/history/www/homepage_archive/2018/october_2018.html.

11. NPR. "The Phoebus Cartel." *Throughline,* March 28, 2019. Podcast MP3, 32:00. www.npr.org/transcripts/707188193.

12. NPR. "The Phoebus Cartel." *Throughline,* March 28, 2019. Podcast MP3, 32:00. www.npr.org/transcripts/707188193.

13. Limer, Eric. "The Stubborn Man Still Using His Nokia 3310 after 17 Years." *Popular Mechanics,* February 13, 2017. www.popularmechanics.com/technology/gadgets/a25192/man-uses-nokia-17-years.

14. Gill, Victoria. "E-waste: Five Billion Phones to Be Thrown Away in 2022." BBC News, October 14, 2022. www.bbc.com/news/science-environment-63245150.

15. Porter, Jon, and James Vincent. "USB-C Will Be Mandatory for Phones Sold in the EU 'By Autumn 2024.'" *The Verge,* June 7, 2022. www.theverge.com/2022/6/7/23156361/european-union-usb-c-wired-charging-iphone-lightning-ewaste.

16. "Oscar's Trash Can." Muppet Wiki. https://muppet.fandom.com/wiki/Oscar%27s_trash_can.

17. Stockholm Environment Institute. *Transformational Change through a Circular Economy.* Stockholm Environment Institute, 2019. www.jstor.org/stable/resrep22978.

Chapter 3: The Beginning of the Garbage Industry

1. Rossi, Marcello. "How Taiwan Has Achieved One of the Highest Recycling Rates in the World." *Smithsonian,* January 3, 2019. www.smithsonianmag.com/innovation/how-taiwan-has-achieved-one-highest-recycling-rates-world-180971150.

2. Ngo, Hope. "How Getting Rid of Dustbins Helped Taiwan Clean Up Its Cities." Future Planet, BBC Future, May 27, 2020. www.bbc.com/future/article/20200526-how-taipei-became-an-unusually-clean-city.

3. United States Environmental Protection Agency. *Recycling Regulations in Taiwan and the 4-in-1 Recycling Program.* Washington, DC: EPA, 2012.

4. Ngo, Hope. "How Getting Rid of Dustbins Helped Taiwan Clean Up Its Cities." Future Planet, BBC Future, May 27, 2020. www.bbc.com/future/article/20200526-how-taipei-became-an-unusually-clean-city.

5. Wilson, Mark. "New York City Has Chosen This Trash Can of the Future." *Fast Company,* December 9, 2019. www.fastcompany.com/90440147/new-york-city-has-chosen-this-trash-can-of-the-future.

6. Jennings, Ralph. "In Taiwan, Leftover Food Scraps Help Farmers Sustain Porky Appetites." *The Guardian,* March 23, 2016. www.theguardian.com/sustainable-business/2016/mar/23/taiwan-food-waste-pork-production-farming-recycling-environment.

7. Johnson, Ben. "The Great Horse Manure Crisis of 1894." Historic UK, July 4, 2017. www.historic-uk.com/HistoryUK/HistoryofBritain/Great-Horse-Manure-Crisis-of-1894.

8. Kohlstedt, Kurt. "The Big Crapple: NYC Transit Pollution from Horse Manure to Horseless

Carriages." *99% Invisible*, November 14, 2017. www.99percentinvisible.org/article/cities-paved -dung-urban-design-great-horse-manure-crisis-1894.

9. Strasser, Susan. *Waste and Want: A Social History of Trash*. New York: Metropolitan Books, 1999.

10. MacBride, Samantha. "The Archeology of Coal Ash: An Industrial-Urban Solid Waste at the Dawn of the Hydrocarbon Economy." *IA. The Journal of the Society for Industrial Archeology* 39, no. 1/2 (2013): 23–39. www.jstor.org/stable/43958425.

11. United States Environmental Protection Agency. "National Overview: Facts and Figures on Materials, Wastes and Recycling." Last modified December 3, 2022. www.epa.gov/facts-and-figures -about-materials-waste-and-recycling/national-overview-facts-and-figures-materials.

12. Ryerson, Jade. "Hull-House and the 'Garbage Ladies' of Chicago." US National Park Service. www.nps.gov/articles/000/hull-house-and-the-garbage-ladies-of-chicago.htm.

13. Cohen, Rich. "The Mobster Who Bought His Son a Hockey Team." *The Atlantic*, May 2018. www.theatlantic.com/magazine/archive/2018/05/the-mobster-who-bought-his-teenage-son-a -hockey-team/556853.

14. Urstadt, Bryant. "Skating with the Mob." *ESPN the Magazine*, October 23, 2006. www.espn.com /espnmag/story?id=3646723.

15. United States Attorney's Office: District of Connecticut. "Former Controller of Danbury Trash Companies Is Sentenced." Archives, The US Federal Bureau of Investigations, August 11, 2009. https://archives.fbi.gov/archives/newhaven/press-releases/2009/nh081109a.htm.

16. Cohen, Rich. "The Mobster Who Bought His Son a Hockey Team." *The Atlantic*, May 2018. www .theatlantic.com/magazine/archive/2018/05/the-mobster-who-bought-his-teenage-son-a-hockey -team/556853.

17. Ascher, Kate. *The Works: Anatomy of a City*. New York: Penguin Press, 2005.

18. Fitzsimmons, Emma G. "Why New York's Giant Trash Bag Piles May Be an Endangered Species." *The New York Times*, May 3, 2023. www.nytimes.com/2023/05/03/nyregion/garbage-containers -nyc-adams.html.

19. Feldman, Kiera. "Trashed: Inside the Deadly World of Private Garbage Collection." ProPublica, January 4, 2018. www.propublica.org/article/trashed-inside-the-deadly-world-of-private-garbage-collection.

20. Feldman, Kiera. "Trashed: Inside the Deadly World of Private Garbage Collection." ProPublica, January 4, 2018. www.propublica.org/article/trashed-inside-the-deadly-world-of-private-garbage -collection.

21. New York City Comptroller. "Unsafe Sanitation: An Analysis of the Commercial Waste Industry's Safety Record." Accessed January 11, 2023. https://comptroller.nyc.gov/reports/unsafe-sanitation -an-analysis-of-the-commercial-waste-industrys-safety-record.

Chapter 4: Poverty vs. Wealth

1. Khan, Themrise, Seye Abimbola, Catherine Kyobutungi, and Madhukar Pai. "How How We Classify Countries and People—and Why It Matters." *BMJ Global Health* 7, no. 6 (2022). http://dx.doi .org/10.1136/bmjgh-2022-009704.

2. Scott, Michon. "Does It Matter How Much the United States Reduces Its Carbon Dioxide Emissions if China Doesn't Do the Same?" Climate.gov, August 30, 2023. www.climate.gov/news -features/climate-qa/does-it-matter-how-much-united-states-reduces-its-carbon-dioxide-emissions.

3. Borenstein, Seth, and Drew Costley. "Rich Nations Caused Climate Harm to Poorer Ones, Study Says." AP News, July 12, 2022. https://apnews.com/article/climate-russia-ukraine-science-united -states-226702e6d195c94433cdc48e5fed6e63.

4. Bruggers, James. "Q&A: Cancer Alley Is Real, and Louisiana Officials Helped Create It, Researchers Find." Inside Climate News, February 8, 2023. https://insideclimatenews.org/news/08022023 /louisiana-cancer-alley.

5. The Climate Reality Project. "Frontline and Fenceline Communities." www
.climaterealityproject.org/frontline-fenceline-communities.

6. Transform Don't Trash NYC. *Clearing the Air: How Reforming the Commercial Waste Sector Can Address Air Quality Issues in Environmental Justice Communities.* New York: Transform Don't Trash NYC, 2016. www.transformdonttrashnyc.org/wp-content/uploads/2016/09/Final-draft-v3_TDT -Air-Qual-Report_Clearing-the-Air-1.pdf.

7. Villarosa, Linda. "Pollution Is Killing Black Americans. This Community Fought Back." *The New York Times Magazine*, July 28, 2020. www.nytimes.com/2020/07/28/magazine/pollution -philadelphia-black-americans.html.

8. Villarosa, Linda. "Pollution Is Killing Black Americans. This Community Fought Back." *The New York Times Magazine*, July 28, 2020. www.nytimes.com/2020/07/28/magazine/pollution -philadelphia-black-americans.html

9. Ellison, Charles D. "Reality Check: One Word to Fix Philly's Trash Crisis? Invest." The Philadelphia Citizen, February 3, 2022. https://thephiladelphiacitizen.org/reality-check-one-word-to-fix-phillys -trash-crisis-invest/.

10. Murphy, Darryl C. "Cleaning Vacant Lots Leads to Safer and Healthier Neighborhoods, New Study Finds." WHYY, February 24, 2018. https://whyy.org/articles/cleaning-vacant-lots-leads-to-safer-and -healthier-neighborhoods-new-study-finds.

11. Boudreau, Catherine. "The Waste Picker Fighting for Global Recognition." *Politico*, April 13, 2022. www.politico.com/newsletters/the-long-game/2022/04/13/the-waste-picker-fighting-for -global-recognition-00024944.

12. The World Bank. "Brief: Solid Waste Management." Last modified September 23, 2019. www .worldbank.org/en/topic/urbandevelopment/brief/solid-waste-management.

13. Laville, Sandra. "Leader of Kenyan Waste Pickers: 'We Are the Backbone of Recycling.'" *The Guardian*, May 12, 2023. www.theguardian.com/environment/2023/may/12/leader-kenya-waste -pickers-we-are-backbone-of-recycling-plastic-pollution.

14. Laville, Sandra. "Leader of Kenyan Waste Pickers: 'We Are the Backbone of Recycling'." *The Guardian*, May 12, 2023. https://www.theguardian.com/environment/2023/may/12/leader-kenya -waste-pickers-we-are-backbone-of-recycling-plastic-pollution.

Chapter 5: Recycling

1. United States Environmental Protection Agency. "Landfill Methane Outreach Program: Basic Information About Landfill Gas." Accessed September 6, 2022. www.epa.gov/lmop/basic -information-about-landfill-gas.

2. Gonzalez, Sarah. "A Mob Boss, a Garbage Boat and Why We Recycle." *Planet Money*, July 10, 2019. Podcast MP3, 25:00. https://www.npr.org/2019/07/09/739893511/episode-925-a-mob-boss-a -garbage-boat-and-why-we-recycle; Hanbury, Harry W. "Retro Report: Voyage of the Mobro 4000." The New York Times, May 21, 2013. YouTube video, 12:07. www.youtube.com/ watch?v=WrugoT8N5cE.

3. Strasser, Susan. *Waste and Want: A Social History of Trash.* New York: Metropolitan Books, 1999.

4. Strasser, Susan. *Waste and Want: A Social History of Trash.* New York: Metropolitan Books, 1999.

5. Busch, Jane. "Second Time Around: A Look at Bottle Reuse." *Historical Archaeology* 21, no. 1 (1987): 67–80. www.jstor.org/stable/25615613.

6. Blair, Peter W., Kevin P. Budris, and Kirstie Pecci. *The Big Beverage Playbook for Avoiding Responsibility.* Boston: Conservation Law Foundation, 2022. www.clf.org/wp-content /uploads/2022/02/2022-02-09-CLF-Beverage-Playbook-Report.pdf.

7. Sure We Can. "Home." www.surewecan.org. Accessed December 7, 2022.

8. Sullivan, Laura. "How Big Oil Misled the Public Into Believing Plastic Would Be Recycled." NPR,

September 11, 2020. www.npr.org/2020/09/11/897692090/how-big-oil-misled-the-public-into -believing-plastic-would-be-recycled.

9.   GrrlScientist. "Festive Fireworks Create Harmful Pall of Pollution." *Forbes*, December 31, 2019. www.forbes.com/sites/grrlscientist/2019/12/31/festive-fireworks-create-harmful-pall-of-pollution/.

10. Leibacher, Herb. "Disney Performs Magic with Trash—Underground Tubes Whisk It Away." *World of Walt* (blog), April 11, 2017. www.worldofwalt.com/disney-underground-trash-tubes.html. (The number represents average attendance before COVID-19 shutdowns.)

11. Chaban, Matt A.V. "Garbage Collection, Without the Noise or the Smell." *The New York Times*, August 3, 2015. www.nytimes.com/2015/08/04/nyregion/garbage-collection-without-the-noise-or -the-smell.html?_r=0.

12. Chaban, Matt A.V. "Garbage Collection, Without the Noise or the Smell." *The New York Times*, August 3, 2015. www.nytimes.com/2015/08/04/nyregion/garbage-collection-without-the-noise-or -the-smell.html?_r=0.

13. Dunaway, Finis. "The 'Crying Indian' Ad That Fooled the Environmental Movement." *Chicago Tribune*, May 31, 2019. www.chicagotribune.com/opinion/commentary/ct-perspec-indian-crying -environment-ads-pollution-1123-20171113-story.html.

14. Sullivan, Laura. "How Big Oil Misled the Public Into Believing Plastic Would Be Recycled." NPR, September 11, 2020. www.npr.org/2020/09/11/897692090/how-big-oil-misled-the-public-into -believing-plastic-would-be-recycled.

Chapter 6: Landfills and Incinerators

1.   Kaza, Silpa, Lisa Yao, Perinaz Bhada-Tata, and Frank Van Woerden. *What a Waste 2.0: A Global Snapshot of Solid Waste Management to 2050.* Urban Development Series. Washington, DC: World Bank, 2018. https://hdl.handle.net/10986/30317.

2.   Rosenthal, Elisabeth. "Europe Finds Clean Energy in Trash, but U.S. Lags." *The New York Times*, April 12, 2010. www.nytimes.com/2010/04/13/science/earth/13trash.html.

3.   National Audit Office Wales. *Environment Agency Wales: Regulation of Waste Management.* London: National Audit Office, 2004. https://web.archive.org/web/20120308220156 /http://www.wao.gov.uk/assets/englishdocuments/Environment_Agency_Wales_Waste _Management_agw_2004.pdf.

4.   United States Environmental Protection Agency. "Superfund Site: Rhinehart Tire Fire Dump Frederick County, VA." Accessed December 4, 2022. https://cumulis.epa.gov/supercpad/SiteProfiles /index.cfm?fuseaction=second.Cleanup&id=0302772#bkground.

5.   Murray, Adrienne. "The Incinerator and the Ski Slope Tackling Waste." *BBC News*, October 4, 2019. www.bbc.com/news/business-49877318.

6.   Schulte, Brigid. "Landfills Report up to 30% Decline." *The Seattle Times*, March 15, 2009. www .seattletimes.com/nation-world/landfills-report-up-to-30-decline.

7.   City of Fresno. "Department of Public Utilities: Solid Waste Facilities." www.fresno.gov /publicutilities/trash-disposal-recycling/solid-waste-facilities.

8.   Sullivan, Robert. "How The World's Largest Garbage Dump Evolved Into a Green Oasis." *The New York Times*, August 14, 2020. Last modified July 23, 2021. www.nytimes.com/2020/08/14/nyregion /freshkills-garbage-dump-nyc.html.

9.   United States Army Corps of Engineers. "9/11 Anniversary Debris Mission." US Army Corps of Engineers, September 2021. www.usace.army.mil/About/History/Historical-Vignettes/Relief-and -Recovery/148-September-11-Debris-Mission.

10. Bram, Jason, James Orr, and Carol Rapaport. "Economic Policy Review Executive Summary: Measuring the Effects of the September 11 Attack on New York City." Federal Reserve Bank of New York, November 2002. www.newyorkfed.org/research/epr/02v08n2/0211rapa/0211rapa.html.

Chapter 7: Food

1. The World Bank. "Trends in Solid Waste Management." *What a Waste 2.0: A Global Snapshot of Solid Waste Management to 2050.* Accessed September 18, 2022. https://datatopics.worldbank .org/what-a-waste/trends_in_solid_waste_management.html.

2. United States Environmental Protection Agency. "National Overview: Facts and Figures on Materials, Wastes and Recycling." Last modified December 3, 2022. www.epa.gov/facts-and-figures -about-materials-waste-and-recycling/national-overview-facts-and-figures-materials.

3. Broad Leib, Emily. *The Dating Game: How Confusing Food Date Labels Lead to Food Waste in America.* New York: Harvard Food Law and Policy Clinic and the Natural Resources Defense Council, 2013. www.nrdc.org/sites/default/files/dating-game-report.pdf.

4. Chandler, Adam. "Why Americans Lead the World in Food Waste." *The Atlantic,* July 15, 2016. www.theatlantic.com/business/archive/2016/07/american-food-waste/491513.

5. Peterson, Shawn. *2020 Gleaning Census.* Salt Lake City: Association of Gleaning Organizations, 2020. nationalgleaningproject.org/wp-content/uploads/2021/11/AGO-2020Census-rev3-web.pdf.

6. Karp, David. "Most of America's Fruit Is Now Imported. Is That a Bad Thing?" *The New York Times,* March 13, 2018. www.nytimes.com/2018/03/13/dining/fruit-vegetables-imports.html.

7. Blakely, Julia. "The Prickly Meanings of the Pineapple." *Unbound* (blog), Smithsonian Libraries and Archives, January 28, 2021. blog.library.si.edu/blog/2021/01/28/the-prickly-meanings-of-the -pineapple.

8. Too Good To Go. *2021 Impact Report.* New York: Too Good To Go, 2021. tgtg-mkt-cms-prod.s3 .eu-west-1.amazonaws.com/27100/USImpactReport.pdf.

Chapter 8: Plastic

1. Miller, Jeremy. "The Plastic in Paint Is Massively Polluting the Ocean." *Sierra Magazine,* February 22, 2022. www.sierraclub.org/sierra/plastic-paint-massively-polluting-ocean.

2. Meikle, Jeffrey. *American Plastic: A Cultural History.* New Brunswick: Rutgers University Press, 1995.

3. CBS Baltimore. "Meet Baltimore's Fourth Trash Wheel: Gwynnda the Good Wheel of the West." CBS News Baltimore, March 11, 2021. www.cbsnews.com/baltimore/news/meet-baltimores-fourth -trash-wheel-gwynnda-the-good-wheel-of-the-west; Mr. Trash Wheel.com. "Home." Accessed December 2, 2022. www.mrtrashwheel.com.

4. Ridder, M. "Coca-Cola Co.: Ad Spend 2014-2021". Statista, September 26, 2023. www.statista.com /statistics/286526/coca-cola-advertising-spending-worldwide.

5. Blair, Peter W., Kevin P. Budris, and Kirstie Pecci. *The Big Beverage Playbook for Avoiding Responsibility.* Boston: Conservation Law Foundation, 2022. www.clf.org/wp-content /uploads/2022/02/2022-02-09-CLF-Beverage-Playbook-Report.pdf.

6. Parker, Laura. "Ocean Trash: 5.25 Trillion Pieces and Counting, but Big Questions Remain." *National Geographic,* January 11, 2015. www.nationalgeographic.com/science/article/150109 -oceans-plastic-sea-trash-science-marine-debris..

7. United States Environmental Protection Agency. "Learn about Ocean Dumping." Accessed October 27, 2022. www.epa.gov/ocean-dumping/learn-about-ocean-dumping.

8. Parker, Laura. "Ocean Trash: 5.25 Trillion Pieces and Counting, but Big Questions Remain." *National Geographic,* January 11, 2015. www.nationalgeographic.com/science/article/150109 -oceans-plastic-sea-trash-science-marine-debris; National Geographic Society. "Great Pacific Garbage Patch." National Geographic Education. Accessed October 28, 2022. https://education. nationalgeographic.org/resource/great-pacific-garbage-patch.

9. National Geographic Society. "Plastic Bag Found at the Bottom of World's Deepest Ocean

Trench." National Geographic Education. Accessed October 28, 2022. https://education. nationalgeographic.org/resource/plastic-bag-found-bottom-worlds-deepest-ocean-trench.

10. Wetzel, Corryn. "The Great Pacific Garbage Patch Hosts Life in the Open Ocean." *Smithsonian*, December 6, 2021. www.smithsonianmag.com/smart-news/the-great-pacific-garbage-patch-hosts -life-in-the-open-ocean-180979168.

11. The Plastiki. "Home." Accessed December 2, 2020. https://theplastiki.com.

12. World Wildlife Foundation Australia. "Plastic in Our Oceans Is Killing Marine Mammals." *World Wildlife Foundation Australia* (blog), June 26, 2023. www.wwf.org.au/news/blogs/plastic-in-our -oceans-is-killing-marine-mammals.

13. Daly, Natasha. "Why Do Ocean Animals Eat Plastic?" *National Geographic,* December 5, 2018. www.nationalgeographic.com/animals/article/whales-eating-plastic-pollution.

14. Bartels, Meghan. "Surprising Creatures Lurk in the Great Pacific Garbage Patch." *Scientific American*, April 17, 2023. www.scientificamerican.com/article/surprising-creatures-lurk-in-the -great-pacific-garbage-patch.

15. United Nations Environment Programme. "Our Planet Is Choking on Plastic." Accessed May 20, 2023. www.unep.org/interactives/beat-plastic-pollution.

16. World Business Council for Sustainable Development. *End-of-Life Tire (ELT) Management Toolkit.* Geneva: World Business Council for Sustainable Development, 2021. www.wbcsd.org /download/file/13232.

17. Florida Department of Environmental Protection. "Osborne Reef Waste Tire Removal Project." Last modified August 15, 2023. https://floridadep.gov/waste/permitting-compliance-assistance/content /osborne-reef-waste-tire-removal-project.

18. Carrington, Damian. "Car Tyres Produce Vastly More Particle Pollution Than Exhausts, Tests Show." *The Guardian*, June 3, 2022. www.theguardian.com/environment/2022/jun/03/car-tyres-produce -more-particle-pollution-than-exhausts-tests-show.

19. Root, Tik. "Tires: The Plastic Polluter You Never Thought About." *National Geographic*, September 20, 2019. www.nationalgeographic.com/environment/article/tires-unseen-plastic-polluter.

20. Carrington, Damian. "Car Tyres Produce Vastly More Particle Pollution Than Exhausts, Tests Show." *The Guardian*, June 3, 2022. www.theguardian.com/environment/2022/jun/03/car-tyres-produce -more-particle-pollution-than-exhausts-tests-show.

21. Carrington, Damien. "Car Tyres Are Major Source of Ocean Microplastics – Study." *The Guardian*, July 14, 2020. https://www.theguardian.com/environment/2020/jul/14/car-tyres-are-major-source -of-ocean-microplastics-study.

22. Lee, Chermaine. "Could Plastic Roads Make for a Smoother Ride?" Future Planet, BBC Future, March 2, 2021. www.bbc.com/future/article/20210302-could-plastic-roads-make-for-a-smoother- ride.

23. Chandran, Rina. "Sturdier, Safer, Cheaper: India Urged to Build More Roads With Plastic Waste." Reuters, October 25, 2017. www.reuters.com/article/us-india-environment-construction/sturdier -safer-cheaper-india-urged-to-build-more-roads-with-plastic-waste-idUSKBN1CU24W.

24. Weir, William. "Asphalt Adds to Air Pollution, Especially on Hot, Sunny Days." *YaleNews*, September 2, 2020. https://news.yale.edu/2020/09/02/asphalt-adds-air-pollution-especially ww-hot-sunny-days.

25. Hunt, Katie. "These Plastic-Chomping Caterpillars Can Help Fight Pollution." CNN, March 4, 2020. www.cnn.com/2020/03/04/world/caterpillars-plastic-scn/index.html.

26. American Society for Microbiology. *FAQ: Microbes and Oil Spills*. Washington, DC: American Society for Microbiology, 2011. https://www.ncbi.nlm.nih.gov/books/NBK562898/.

27. Sharp, Trudy, Andrew Lothian, Adam Munn, and Glen Saunders. *CAN001 Methods for the Field Euthanasia of Cane Toads*. Canberra: Australian Government: Department of Climate Change,

Energy, the Environment and Water, 2011. www.dcceew.gov.au/environment/invasive-species/publications/can001-methods-field-euthanasia-cane-toads.

Chapter 9: Paper

1. Borden, Jeremy. "Searching for McClatchy'S North Carolina Future." The Assembly NC, June 17, 2017. www.theassemblync.com/media/searching-for-mcclatchy-north-carolina-future.

2. Grundy, Adam. "Service Annual Survey Shows Continuing Decline in Print Publishing Revenue." America Counts Stories (blog), United States Census Bureau, June 07, 2022. www.census.gov/library/stories/2022/06/internet-crushes-traditional-media.html.

3. Simonetti, Isabella. "Over 360 Newspapers Have Closed Since Just Before the Start of the Pandemic." The New York Times, June 29, 2022. www.nytimes.com/2022/06/29/business/media/local-newspapers-pandemic.html.

4. United States Environmental Protection Agency. "National Overview: Facts and Figures on Materials, Wastes and Recycling." Last modified December 3, 2022. www.epa.gov/facts-and-figures-about-materials-waste-and-recycling/national-overview-facts-and-figures-materials.

5. Paben, Jared. "Analysts: US OCC recycling rate may be below 70%." Resource Recycling, updated September 2, 2022. www.resource-recycling.com/recycling/2022/08/29/analysts-us-occ-recycling-rate-may-be-below-70/.

6. Californians Against Waste. "Skip the Slip: Paper Receipts on Request." www.cawrecycles.org/ab-161-ting.

7. Wishart, Lucy. "Amazon Is Destroying Millions of Unsold Goods, From Smart TVs to Laptops." Fast Company, July 3, 2021. www.fastcompany.com/90652617/amazon-is-destroying-millions-of-unsold-goods-from-smart-tvs-to-laptops.

8. Clark, Alasdair. "Amazon Dunfermline Probe Suggests 'Millions' of Unsold Items Destroyed." The Courier (UK), June 21, 2021. www.thecourier.co.uk/fp/news/fife/2325205/amazon-dunfermline-probe-suggests-millions-of-unsold-items-being-destroyed.

9. Hoffower, Hillary. "We Did the Math to Calculate How Much Money Jeff Bezos Makes in a Year, Month, Week, Day, Hour, Minute, and Second." Business Insider, January 13, 2020. www.businessinsider.com/what-amazon-ceo-jeff-bezos-makes-every-day-hour-minute-2018-10.

10. Oceana. The Cost of Amazon's Plastic Denial on the World's Oceans. Washington, DC: Oceana, 2022, https://oceana.org/reports/the-cost-of-amazons-plastic-denial.

11. Weise, Karen. "Amazon's Profit Soars 220 Percent as Pandemic Drives Shopping Online." The New York Times, May 12, 2021. www.nytimes.com/2021/04/29/technology/amazons-profits-triple.html.

12. Nguyen, Nicole. "The Hidden Environmental Cost of Amazon Prime's Free, Fast Shipping." BuzzFeed News, July 21, 2018. https://www.buzzfeednews.com/article/nicolenguyen/environmental-impact-of-amazon-prime.

Chapter 10: Textiles

1. United States Environmental Protection Agency. "Textiles: Material-Specific Data." Last modified December 3, 2022. www.epa.gov/facts-and-figures-about materials-waste-and-recycling/textiles-material-specific-data.

2. Pucker, Kenneth P. "The Myth of Sustainable Fashion." Harvard Business Review, January 13, 2022. https://hbr.org/2022/01/the-myth-of-sustainable-fashion.

3. Bartlett, John. "Fast Fashion Goes To Die in the World's Largest Fog Desert. The Scale Is Breathtaking." National Geographic, April 10, 2023. www.nationalgeographic.com/environment/article/chile-fashion-pollution?loggedin=true&rnd=1684601367359.

4. Strasser, Susan. Waste and Want: A Social History of Trash. New York: Metropolitan Books, 1999.

5. The Editors of Encyclopaedia Britannica. "Rayon." Britannica.com, April 4, 2016. www.britannica.com/technology/rayon-textile-fibre.

6. Cutlip, Kimbra. "How Nylon Stockings Changed the World." *Smithsonian*, May 11 2015. www
    .smithsonianmag.com/smithsonian-institution/how-nylon-stockings-changed-world-180955219.

7. Leach, William. *Land of Desire: Merchants, Power, and the Rise of a New American Culture*. New
    York: Vintage Books, 1994.

8. European Environment Agency. *Plastic in Textiles: Towards a Circular Economy for Synthetic
    Textiles in Europe*. European Environment Agency, 2021. https://doi.org/10.2800/555165.

9. Cowley, Jenny, Charlsie Agro, and Stephanie Matteis. "Experts Warn of High Levels of Chemicals in
    Clothes by Some Fast-fashion Retailers." CBC News, October 1, 2021. www.cbc.ca/news/business
    /marketplace-fast-fashion-chemicals-1.6193385.

10. Pucker, Kenneth P. "The Myth of Sustainable Fashion." *Harvard Business Review*, January 13, 2022.
    https://hbr.org/2022/01/the-myth-of-sustainable-fashion.

11. Al Jazeera. "Chile's Desert Dumping Ground for Fast Fashion Leftovers." Al Jazeera, November 8,
    2021. www.aljazeera.com/gallery/2021/11/8/chiles-desert-dumping-ground-for-fast-fashion-leftovers.

12. NPR. "PEOPLE: Planet Money Makes a T-Shirt (part 3)." NPR. Video, 6:21. https://apps.npr.org
    /tshirt/#/people.

13. Paddison, Laura. "Single Clothes Wash May Release 700,000 Microplastic Fibres, Study Finds." *The
    Guardian*, September 26, 2016. www.theguardian.com/science/2016/sep/27/washing-clothes
    -releases-water-polluting-fibres-study-finds.

14. Tonti, Lucianne. "How Green Are Your Leggings? Recycled Polyester Is Not a Silver Bullet (Yet)."
    The Guardian, March 21, 2021. www.theguardian.com/fashion/2021/mar/22/how-green-are-your
    -leggings-recycled-polyester-is-not-a-silver-bullet-yet.

15. Waxman, Olivia. "People Have Been Reusing Clothes Forever but Thrift Shops Are Relatively New.
    Here's Why." *Time*, August 17, 2018. https://time.com/5364170/thrift-store-history/.

16. Mull, Amanda. "Seriously, What Are You Supposed to Do With Old Clothes?" *The Atlantic*, August 3,
    2022. www.theatlantic.com/technology/archive/2022/08/what-to-do-with-old-clothing-donation
    -waste/671043.

17. Bartlett, John. "Fast Fashion Goes To Die in the World's Largest Fog Desert. The Scale Is
    Breathtaking." *National Geographic*, April 10, 2023. www.nationalgeographic.com/environment
    /article/chile-fashion-pollution?loggedin=true&rnd=1684601367359.

18. BBC News. "Burberry Burns Bags, Clothes and Perfume Worth Millions." BBC News, July 19, 2018.
    www.bbc.com/news/business-44885983.

19. Dwyer, Jim. "Slashers' Work Ruins Shoes Discarded at a Nike Store." *The New York Times*, January
    26, 2017. www.nytimes.com/2017/01/26/nyregion/slashers-work-ruins-shoes-discarded-at-a-nike
    -store.html.

20. Dweyr, Jim. "Trademark Trumps Charity, so U.S. Will Destroy Bogus N.F.L. Jerseys." *The New York
    Times*, January 30, 2014. www.nytimes.com/2014/01/31/nyregion/trademark-trumps-charity-so-us
    -will-destroy-bogus-nfl-jerseys.html.

21. Dwyer, Jim. "Closing Pipeline to Needy, City Shreds Clothes." *The New York Times*, January 12, 2010.
    www.nytimes.com/2010/01/13/nyregion/13about.html.

22. Cline, Elizabeth L. "Most Fashion Brands Don't Know Enough About Their Carbon Footprints to
    Actually Shrink Them." *Fashionista*, October 25, 2019. https://fashionista.com/2019/10/fashion
    -brands-carbon-fooprints-reducing-emissions.

Chapter 11: Construction

1. Purchase, Callun Keith, Dhafer Manna Al Zulayq, Bio Talakatoa O'Brien, Matthew Joseph
    Kowalewski, Aydin Berenjian, Amir Hossein Tarighaleslami, and Mostafa Seifan. "Circular
    Economy of Construction and Demolition Waste: A Literature Review on Lessons, Challenges, and
    Benefits." *Materials* 15, no. 1 (2021): 76. https://doi.org/10.3390/ma15010076.

2. United States Environmental Protection Agency. "Sustainable Marketplace: Greener Products and Services, Identifying Greener Carpet." Last modified June 14, 2022. www.epa.gov/greenerproducts /identifying-greener-carpet.

3. Williams, Fran. "Virtuous Circles: Can Reusing Building Materials in New Projects Go Mainstream?" *The Architects' Journal,* January 17, 2020. www.architectsjournal.co.uk/news/virtuous-circles-can -reusing-building-materials-in-new-projects-go-mainstream.

4. Purchase, Callun Keith, Dhafer Manna Al Zulayq, Bio Talakatoa O'Brien, Matthew Joseph Kowalewski, Aydin Berenjian, Amir Hossein Tarighaleslami, and Mostafa Seifan. "Circular Economy of Construction and Demolition Waste: A Literature Review on Lessons, Challenges, and Benefits." *Materials* 15, no. 1 (2021): 76. https://doi.org/10.3390/ma15010076.

5. Lendager. "Projects: Resource Rows," Accessed December 10, 2022. http://lendager.com/project /resource-rows/

Chapter 12: Mining

1. Earthworks. "Toxic Release Inventory—What Is It?" Accessed June 21, 2022. https://earthworks.org /issues/toxics-release-inventory-what-is-it/

2. Brown, Matthew. "50m Gallons of Polluted Water Pours Daily from U.S. Mine Sites." Associated Press, February 20, 2019.

3. Morgenstern, Norbert R. "Geotechnics and Mine Waste Management—Update." *EurekaMag,* 1998. https://eurekamag.com/research/019/075/019075728.php

4. Tran, Lina. "Extreme Weather Is Making Mining Waste a Major Problem." Greenbiz, June 10, 2022. www.greenbiz.com/article/extreme-weather-making-mining-waste-major-problem.

5. Early, Catherine. "The New 'Gold Rush' for Green Lithium." Future Planet, BBC Future, November 24, 2020. www.bbc.com/future/article/20201124-how-geothermal-lithium-could-revolutionise-green -energy.

6. International Energy Agency. *The Role of Critical Minerals in Clean Energy Transitions.* Paris: IEA, 2021. www.iea.org/reports/the-role-of-critical-minerals-in-clean-energy-transitions/executive -summary.

7. Gross, Terry. "How 'Modern-Day Slavery' in the Congo Powers the Rechargeable Battery Economy." *Goats and Soda* (blog), NPR, February 1, 2023. www.npr.org/sections /goatsandsoda/2023/02/01/1152893248/red-cobalt-congo-drc-mining-siddharth-kara.

8. Gross, Terry. "How 'Modern-Day Slavery' in the Congo Powers the Rechargeable Battery Economy." *Goats and Soda* (blog), NPR, February 1, 2023. www.npr.org/sections /goatsandsoda/2023/02/01/1152893248/red-cobalt-congo-drc-mining-siddharth-kara.

9. Dietrich Brauch, Martin, Mara Greenberg, Tehtana Mebratu-Tsegaye, and Perrine Toledano. *Five Years after the Adoption of the Paris Agreement, Are Climate Change Considerations Reflected in Mining Contracts?* New York: Columbia Center on Sustainable Development, 2021, https:// ccsi.columbia.edu/sites/default/files/content/docs/ccsi-climate-change-investor-state-mining -contracts.pdf.

10. Delevingne, Lindsey, Will Glazener, Liesbet Grégoir, and Kimberly Henderson. *Climate Risk and Decarbonization: What Every Mining CEO Needs to Know.* McKinsey Sustainability, January 28, 2020. www.mckinsey.com/capabilities/sustainability/our-insights/climate-risk-and-decarbonization -what-every-mining-ceo-needs-to-know.

Chapter 13: Radioactive Waste

1. International Atomic Energy Agency. "Frequently Asked Chernobyl Questions." www .iaea.org/newscenter/focus/chernobyl/faqs.

2. Chaisson, Clara. "Fossil Fuel Air Pollution Kills One in Five People." National Resources Defense Council, February 19, 2021. www.nrdc.org/stories/fossil-fuel-air-pollution-kills-one-five-people.

3. Harvard T. H. Chan School of Public Health. "Fossil Fuel Air Pollution Responsible for 1 in 5 Deaths Worldwide." C-Change: Center for Climate, Health, and the Global Environment, February 9, 2021. www.hsph.harvard.edu/c-change/news/fossil-fuel-air-pollution-responsible-for-1-in-5-deaths -worldwide.

4. El-showk, Sedeer. "Final Resting Place." *Science*, February 24, 2022. www.science.org/content/article /finland-built-tomb-store-nuclear-waste-can-it-survive-100000.

Chapter 14: E-waste

1. International Energy Agency. "Data Centres and Data Transmission Networks." www.iea.org/energy -system/buildings/data-centres-and-data-transmission-networks.

2. Ceci, Laura. "Number of E-mails per Day Worldwide 2017-2026." Statista, August 22, 2023. www .statista.com/statistics/456500/daily-number-of-e-mails-worldwide.

3. Vopson, Melvin M. "The World's Data Explained: How Much We're Producing and Where It's All Stored." *The Conversation*, May 4, 2021. https://theconversation.com/the-worlds-data-explained -how-much-were-producing-and-where-its-all-stored-159964.

4. BBC News. "Bitcoin Mining Producing Tonnes of Waste." BBC News, September 20, 2021. www.bbc .com/news/technology-58572385.

5. Khattak, Rozina. "The Environmental Impact of E-Waste." Earth.Org, March 13, 2023. https://earth .org/environmental-impact-of-e-waste.

6. United States Department of Justice. *A Review of Federal Prison Industries' Electronic-Waste Recycling Program*. Washington, DC: US Department of Justice, 2007. https://oig.justice.gov/reports/BOP /o1010.pdf.

7. Singh, Ana. "Out of Sight, Out of Mind: How the United States Discards E-Waste." *Berkeley Political Review*. December 5, 2019. https://bpr.berkeley.edu/2019/12/05/out-of-sight-out-of-mind-how-the -united-states-discards-e-waste.

8. Larmer, Brook. "E-Waste Offers an Economic Opportunity as Well as Toxicity." *The New York Times*, July 5, 2018. www.nytimes.com/2018/07/05/magazine/e-waste-offers-an-economic-opportunity-as -well-as-toxicity.html.

Chapter 15: Medical Waste

1. Maddipatla, Manojna and Emma Farge. "Huge Volumes of COVID Hospital Waste Threaten Health— WHO." Reuters, Feb 1, 2022. www.reuters.com/business/healthcare-pharmaceuticals/huge-volumes -covid-hospital-waste-threaten-health-who-2022-02-01.

2. Benson, Nsikak U., David E. Bassey, and Thavamani Palanisami. "COVID Pollution: Impact of COVID-19 Pandemic on Global Plastic Waste Footprint." *Heliyon* 7, no. 2 (2021): e06343. https://doi .org/10.1016/j.heliyon.2021.e06343.

3. Korducki, Kelli María. "The World Is Throwing Away 3 Million Face Masks Every Minute—and the Growing Mountain of Waste Is a Toxic Time Bomb." *Business Insider*, December 14, 2022. www .businessinsider.com/disposable-face-mask-waste-toxic-poison-water-nanoplastics-pollution-health -2022-2.

4. Lupkin, Sydney, and Tamara Keith. "Biden Promised a Billion COVID Tests. Contracts to Buy Them Are Being Announced." NPR, January 14, 2022. www.npr.org/2022/01/14/1073215447/biden -promised-a-billion-covid-tests-contracts-to-buy-them-are-being-announced.

5. Our World in Data. "Coronavirus (COVID-19) Vaccinations." https://ourworldindata.org/covid-vaccinations.

6. World Health Organization. "WHO (COVID-19) Home Page: The United Kingdom Situation." https://covid19.who.int/region/euro/country/gb.

7. Jain, Navami, and Desiree LaBeaud. "How Should US Health Care Lead Global Change in Plastic Waste Disposal?" *AMA Journal of Ethics* 24, no. 10 (2022): E986–93. https://doi.org/10.1001 /amajethics.2022.986.

8. Allen, Marshall. "What Hospitals Waste." ProPublica, March 9, 2017, www.propublica.org/article/what-hospitals-waste.

9. Windfeld, Elliott Steen, and Marianne Su-Ling Brooks. "Medical Waste Management—a Review." *Journal of Environmental Management* 163 (2015): 98–108. https://doi.org/10.1016/j.jenvman.2015.08.013.

10. Ngo, Hope. "How Do You Fix Healthcare's Medical Waste Problem?" Future Planet, BBC Future, August 13, 2020. www.bbc.com/future/article/20200813-the-hidden-harm-of-medical-plastic-waste-and-pollution.

11. Allen, Marshall. "The Myth of Drug Expiration Dates." ProPublica, July 18, 2017. www.propublica.org/article/the-myth-of-drug-expiration-dates.

12. Allen, Marshall. "The Myth of Drug Expiration Dates." ProPublica, July 18, 2017. www.propublica.org/article/the-myth-of-drug-expiration-dates.

13. Allen, Marshall. "America's Other Drug Problem." ProPublica, April 27, 2017. www.propublica.org/article/americas-other-drug-problem.

14. Allen, Marshall. "What Hospitals Waste." ProPublica, March 9, 2017. www.propublica.org/article/what-hospitals-waste.

Chapter 16: Human Waste, Sewer Systems, and More

1. UNESCO World Water Assessment Programme. *The United Nations World Water Development Report 2017*. Paris: UNESCO Digital Library, 2017. https://unesdoc.unesco.org/ark:/48223/pf0000247552.

2. Chris Clapp, in discussion with the author, November 2022.

3. BBC News. "Wet Wipe Pollution: 'Fine to Flush' Message Still Not Understood." BBC News, June 24, 2022. bbc.com/news/newsbeat-61922999.

4. Chris Clapp, in discussion with the author, November 2022.

5. Shaver, Katherine. "A Nasty Pandemic Problem: More Flushed Wipes Are Clogging Pipes, Sending Sewage Into Homes." *The Washington Post*, April 23, 2021. https://washingtonpost.com/local/trafficandcommuting/flushable-wipes-clogging-sewers/2021/04/23/5e8bbc82-a2c9-11eb-a774-7b47ceb36ee8_story.html.

6. Moss, Stephen. "Don't Feed The Fatberg! What a Slice of Oily Sewage Says about Modern Life." *The Guardian*, February 18, 2018. www.theguardian.com/environment/2018/feb/18/dont-feed-fatberg-museum-london-clogging-sewers-oil.

7. Moss, Stephen. "Don't Feed The Fatberg! What a Slice of Oily Sewage Says about Modern Life." *The Guardian*, February 18, 2018. www.theguardian.com/environment/2018/feb/18/dont-feed-fatberg-museum-london-clogging-sewers-oil.

8. Adee, Sally. "Myth Busted: Dumped Pills Aren't Main Source of Drugs in Sewage." *New Scientist*, September 9, 2016. www.newscientist.com/article/2105426-myth-busted-dumped-pills-arent-main-source-of-drugs-in-sewage.

9. Safford, Hannah R., Karen Shapiro, and Heather N. Bischel. "Wastewater Analysis Can Be a Powerful Public Health Tool—if It's Done Sensibly." *Proceedings of the National Academy of Sciences* 119, no. 6 (2022). https://doi.org/10.1073/pnas.2119600119.

10. United Press International. "The Last Episode of 'M-A-S-H' Was a Royal Flush . . ." United Press International, March 14, 1983. www.upi.com/Archives/1983/03/14/The-last-episode-of-M-A-S-H-was-a-royal-flush/3459416466000.

11. Ellen MacArthur Foundation. *A Circular Economy for Nappies*. Isle of Wight (UK): Ellen MacArthur Foundation, 2020. bbia.org.uk/wp-content/uploads/2020/11/A-Circular-Economy-for-Nappies-final-oct-2020.pdf.

12. Chris Clapp, in discussion with the author, November 2022.

Chapter 17: The Funeral Industry

1. Ray, Shantanu Guha. "In Varanasi, a Lifetime Spent in a World of Death." *India Ink*, March 16, 2014. https://archive.nytimes.com/india.blogs.nytimes.com/2014/03/16/in-varanasi-a-lifetime-spent-in-a-world-of-death.

2. McBride, Pete. "The Pyres of Varanasi: Breaking the Cycle of Death and Rebirth." *National Geographic*. August 7, 2014. www.nationalgeographic.com/photography/article/the-pyres-of-varanasi-breaking-the-cycle-of-death-and-rebirth.

3. Kermeliotis, Teo. "India's Burning Issue With Emissions From Hindu Funeral Pyres." CNN, September 17, 2011. https://edition.cnn.com/2011/09/12/world/asia/india-funeral-pyres-emissions.

4. Chiappelli, Jeremiah, and Ted Chiappelli. "Drinking Grandma: The Problem of Embalming." *Journal of Environmental Health* 71, no. 5 (2008): 24–29. www.jstor.org/stable/26327817.

5. Green Burial Council. "Disposition Statistics." www.greenburialcouncil.org/media_packet.html.

Chapter 18: Orbital Debris

1. Boley, Aaron C., and Michael Byers. "Satellite Mega-constellations Create Risks in Low Earth Orbit, the Atmosphere and on Earth." *Scientific Reports* 11 (2021): 10642. https://doi.org/10.1038/s41598-021-89909-7.

2. Pultarova, Tereza. "SpaceX Starlink Satellites Responsible for Over Half of Close Encounters in Orbit, Scientist Says." Space.com, August 20, 2021. www.space.com/spacex-starlink-satellite-collision-alerts-on-the-rise.

3. NASA. Report IG-21-011. *NASA's Efforts to Mitigate the Risks Posed by Orbital Debris*. NASA Office of Inspector General, January 27, 2021. https://oig.nasa.gov/docs/IG-21-011.pdf.

4. NASA. Report IG-21-011. *NASA's Efforts to Mitigate the Risks Posed by Orbital Debris*. NASA Office of Inspector General, January 27, 2021. https://oig.nasa.gov/docs/IG-21-011.pdf.

5. Kilic, Cagri. "Mars Is Littered With 15,694 Pounds of Human Trash From 50 Years of Robotic Exploration." Space.com, September 28, 2022. www.space.com/mars-littered-with-human-trash.

6. Meinzer, Kristen. "Galactic Garbage Can: There's 400,000 Pounds of Trash on the Moon." WNYC Studios, January 14, 2015. www.wnycstudios.org/podcasts/takeaway/segments/trash-moon-and-why-its-there.

7. Garber, Megan. "The Trash We've Left on the Moon." *The Atlantic*, December 19, 2012. www.theatlantic.com/technology/archive/2012/12/the-trash-weve-left-on-the-moon/266465.